Sharing Maths Cultures:

IMPACT
Inventing Maths for Parents and Children and Teachers

Ruth Merttens
Jeff Vass

UK	The Falmer Press, Rankine Road, Basingstoke, Hampshire, RG24 0PR
USA	The Falmer Press, Taylor & Francis Inc., 1900 Frost Road, Suite 101, Bristol, PA 19007

© Ruth Merttens and Jeff Vass 1990

All rights reserved. No part of this publication may be reproduced, stored in a retrieval system, or transmitted, in any form or by any means, electronic, mechanical, photocopying, recording or otherwise, without permission in writing from the copyright holder and the Publisher.

First published 1990

British Library Cataloguing in Publication Data
Merttens, Ruth
  Sharing maths culture : IMPACT : Inventing Maths for Parents and Children and Teachers.
  1. Great Britain. Schools. Curriculum subjects : Mathematics.   Teaching. Cooperation between schools & parents
  I. Title   II. Vass, Jeff
  510.71041

  ISBN 1-85000-875-2
  ISBN 1-85000-876-0 pbk

Library of Congress Cataloging-in-Publication Data is available on request

Printed in Great Britain by Burgess Science Press, Basingstoke on paper which has a specified pH value on final paper manufacture of not less than 7.5 and is therefore 'acid free'.

Sharing Maths Cultures

LIBRARY
WESTMINSTER COLLEGE
OXFORD, OX2 9AT

110624

Contents

Preface		vii
Chapter 1	**Introduction: Debate, History and Challenge**	1
	Curriculum, Community and Relevance	2
	Years of Social and Professional Change	4
	A Brief History of IMPACT	11
	An Intervention Project	13
	A Research Project	15
	The IMPACT Network	18
Chapter 2	**Children as Tutors**	21
	What is IMPACT?	21
	The Social Context of Maths	25
	Cooperative Learning	29
	Peer Tutoring	34
	Home and School as Situations for Learning	35
	The IMPACT Process	36
	IMPACT and the National Curriculum	39
Chapter 3	**Getting Started**	42
	National Curriculum Process	42
	IMPACT and School Policy	43
	Staffroom Decisions — Things to Consider	46
	Informing the Governors	53
Chapter 4	**IMPACT in the Classroom and the Home**	57
	Planning IMPACT	57
	Linking the Classwork to the Home Maths	60
	The IMPACT Activity Cycle	70
	Assessment and Record-keeping	76

	Listening to Parents	81
	Patterns of Parental Contact	98
Chapter 5	**Designing and Selecting IMPACT Materials**	105
	Designing IMPACT Activities	106
	Selecting IMPACT Activities	116
	IMPACT Flexibility	127
Chapter 6	**Evaluation: Reflecting on Practice**	140
	Re-thinking the Scope of Evaluation	140
	Evaluation in Complexity	142
	Knowing What to Evaluate	142
	Rationale for Monitoring	144
	Starting with Intuitions	146
	Monitoring IMPACT in the School	147
	Using Evaluation for Critical Reviews of Practice	155
	National Assessment and IMPACT: an Interim Discussion	158

Afterword   161

Appendix with Sample Materials   168

Notes and Bibliography   193

Preface

IMPACT, the project, is about involving parents in the school curriculum through the 'tutelage' of their children and through sustained patterns of direct contact. This book is a distillation of our experience implementing IMPACT firstly in three and then twenty Local Education Authorities.

When the IMPACT team collect their mail in the mornings it is certain to contain letters from teachers, student teachers, lecturers and parents all wanting to know more about IMPACT: 'How does it work?; Is it like shared reading?; What sort of materials do you use?; Is it a form of peer tutoring?; How many parents participate?; Is there research?; How does it fit with the National Curriculum?; What is the theoretical framework? . . .', and so they go on.

This book is addressed to all you patient people who have shown considerable interest over the last few years. It is also addressed to those who, picking up the book, encounter 'IMPACT' for the first time. At the time of writing there are 600 schools affiliated to the IMPACT National Network which we currently coordinate from London. But we know that there are many more schools that have been involving parents in the primary curriculum in ways similar to ourselves. We feel that the time is now ripe to pool ideas, strategies, frameworks and experiences.

We hope this book raises as many questions as it answers. The image of a child at home with a parent working together on a curriculum-based task is a deceptively simple one. One hardly has time to ask 'how?' and 'with what?' before voicing 'why?' and 'to what effect?'. And that is how it should be. These are the questions we tackle in this book. The primary purpose, however, is to provide sufficient information for the reader to try out IMPACT in *their* school.

Chapter 1 locates IMPACT in the on-going debates in education, particularly the relationship of the community to the curriculum. We outline the historical background to IMPACT and place it in its legislative context. Chapter 2 takes a more detailed look at the research context of IMPACT. We cannot proceed in our quest to gain deeper practical insights into the bases of learning and development until we construct an articulate theoretical framework for doing so. The authors of this book feel that the makings of such a framework do exist. We outline parts of this framework in chapter 2. It is not necessary to grasp everything laid out here on a first reading. And it is not necessary to understand this material to 'do' IMPACT. Perhaps after having set up and tried IMPACT strategies you might want to come back and have another look at the parts of chapter 2 that you missed the first time round. To appreciate IMPACT means appreciating the process of learning and development that we find ourselves working with when children themselves become the *tutors* in collaborative work. Most of all we hope you will follow up some of the references given for further reading.

Chapter 3 gives you all you need to know to set up IMPACT in your school. It anticipates the practical and social difficulties you may experience and suggests strategies to overcome them. Chapter 4 concentrates on the classroom and looks at the day-to-day running of IMPACT in school. We focus on fitting IMPACT into your planning and record-keeping. IMPACT, you will discover, suits most ways of planning and implementing the National Curriculum. There are numerous examples given along with other ideas that we have found to work well.

Chapter 5 looks at designing and selecting appropriate materials which extend classwork so that suitable tasks can be done at home jointly with a parent. Ways of organizing these materials are discussed along with strategies for keeping up the momentum.

Chapter 6 takes a look at the issues and techniques of 'complex evaluation' that an intervention like IMPACT requires. School life is complex and IMPACT entails qualitative changes to that life. Such change is not uniform throughout the school let alone the same in every school. We discuss an ongoing evaluation and review process that becomes part of routine practice and part of the new relationships that are engendered with a school's parents.

Finally, we hope to convey a sense of work in progress rather than to give answers with an air of accomplished finality. There are so many people who come to mind as we think of the work of half a decade that has seen so many changes in education. To name everyone would fill many volumes, but we should mention David Coulby, Alistair Ross, Christine Graham, Kate Frood, and Nicky Hayden from the halcyon days. We could not have done without the support of Greg Condry, Peter Huckstep, David Bristow and Anna Morris, and all our new friends on the IMPACT Network. Conversations with Dorothy Hamilton, Dave Wood and John Shotter, and breakfasts with Bill Laar and Eric Greer have been invaluable. We would have collapsed without Deborah Curle, Ian Merttens, Pat (don't panic) Brown and Ros Leather. The support and friendship of Richard Border transformed our experience of the project.

This book is dedicated to all the members of the IMPACT community, and especially to the pilot schools in ILEA, Oxfordshire, Redbridge and Barnet.

Chapter 1

Introduction: Debate, History and Challenge

I would rather my son learnt to speak in the taverns than in the talking-schools.

Michel de Montaigne

In a book which sets out to describe in some detail a do-it-yourself IMPACT, we find we have to start off with a series of questions. These are not rhetorical questions — we simply do not know all the answers. IMPACT is a project that has turned heads in Australia, Eire, Scandinavia, Germany and the USA. Educationatists from far and wide, engaged in their own work, look at IMPACT for the same reason we look at theirs. All of us are part of the world of teaching and learning and are already doing things. You cannot first design projects in the abstract and then see how they will fit with what is going on in the world. This is partly because the world we inhabit ceaselessly poses us with new problems, challenges and opportunities to which we must respond. This means that in the middle of creating initiatives, responding and improvizing we have to talk and question. There are no 'once and for all' answers or strategies.

We attend numerous meetings in schools which 'do' IMPACT. Teachers and parents ask fundamental questions at these gatherings about the nature of education, about why we do what we do as well as the highly specific questions about skills, numbers, maths language and those things which relate to more local issues. Plunging head first into some of these questions will introduce IMPACT and reveal some of education's most dominant themes: themes that transcend the narrower confines of a project that involves parents in the maths curriculum.

Some Questions

Have you ever wondered why we teach long division when we never appear to use it — yet become worried if our own children are not taught it? Do you panic about the maths you have to teach? Are you puzzled by anxious parents who cannot see any 'maths' in school work? Are you confident about engaging in debates about the place of sums and methods of teaching tables in the maths curriculum? Do you find advice from parents, advisors, inspectors and even the literature conflicting? All these questions come at us in the middle of a situation where we are already having to teach.

To outsiders it often appears that these issues break down into two kinds: *what* we teach and *how* we teach. It then seems relatively simple to say that *how* to teach is the business of schools whereas *what* to teach is the business of the wider community. We tend to feel that somewhere in the wider community are the reasons why we teach *what* we teach. Somehow the world 'outside' school will suggest what a 'relevant' curriculum might look like. 'Inside' school, it is often said, we have a better grasp of *how* to deliver that curriculum.

From a classroom perspective, however, such a categorization seems simple-minded. Take a maths example to illustrate the blurring of *what* and *how* in practice. It is possible to treat subtraction as a form of addition (subtracting by adding on). If you do this you will be committed to a style of teaching, a method of *how*, which is quite different to those styles which treat addition and subtraction, and other *whats*, as separate mathematical 'topics'. Some may see us failing to deliver what the community needs if children do not leave school able to perform subtractions in specific ways which the community traditionally recognizes as 'proper' subtraction. The question of relating the curriculum to the vicissitudes of teaching, to the development of children and to the community at large, poses far more problems for schools than one might suppose.

Curriculum, Community and Relevance

The question of relevance arises repeatedly in these discussions. We, apparently, need to discuss the relevance of, say, long division in its contribution to other aspects of mathematics within the curriculum. But it appears we often get into situations where we have to account for bits of the curriculum, say, 'tangrams' to those whom we perceive as outsiders. And this is hard. We are prone to think that their grasp of the curriculum is inadequate. Perhaps what we often mean here is that they do not understand the context of teaching in the classroom.

In these circumstances people lose sight of their interests and lose sight of what they want education to be. We know that relevance is important but we are not clear about recognizing it and communicating it. We often feel that it must have been easier in the past. We might now laugh at the idea that intelligent people were only those who could conjugate Latin verbs and read Greek. But we might secretly envy an age in which the *what* and *how* of the curriculum were comfortably settled. It may not always have been clear where Latin and Greek were relevant to society at large — but they defined education very well and most people seemed to agree, both inside and outside schools. The disconcerting thing is that even today we are as likely to misrecognize (and mis-construct) intelligence as we ever were.

Introduction: Debate, History and Challenge

Perhaps we need to focus more on the pluralism in our experience and understanding of the curriculum.

Was it less pluralistic in the fifth century BC? Music was central to the curriculum that Plato discusses in ancient Athens (*see The Republic*). Music was central because it was conceived as being useful to society. It helped, among other things, in the training of warriors. Plato perceived the intimate relationship between the curriculum and the community at large,

> Any alteration in the modes of music is always followed by alteration in the most fundamental laws of the state. (Book IV, *The Republic*)[1]

Who would place music so centrally in the curriculum now? What agreement would there be? The issue may seem far fetched and somewhat academic. Perhaps we are being romantic, but we often find something compelling in the idea that the social, religious, educational, artistic and military endeavours of a people should have common and communal viewpoints. It is from such viewpoints that the democratic individual is able to articulate their opinions intelligently in each sphere of life. Maybe we should not look for Golden Ages in the distant past; there is no indication they actually existed. We now live 'compartmentalized' lives. And we invest much energy in keeping the compartments separate. We could be a parent and a teacher and be two 'different people' in the one body.

Compartments or fragments are not necessarily bad things if we are able to use them to create new situations for ourselves. But with each of us, for 'professional' or 'natural' reasons, having access only to parts, being 'just' teachers, 'just' parents or 'just' milkmen, we lack ways of entering other parts of the common world which, nevertheless, exert great influence on our lives. We behave as if the parts we have and the roles we occupy are naturally given and there is no further discussion, 'it's just the way things are'. Looking again at Plato's dialogues we have to say that we have not finished the discussion he begins in *The Republic*.

The progress of the 1988 Education Reform Act (ERA) became a media spectacle in which the theme of relevance re-surfaced as continuous banter on numerous carnival floats. Relevance became, on some lurid displays, simply 'what parents wanted'; on others more sombre it was 'what industry needed'. Teachers watched the *mardi gras* with apprehension knowing they would have to do the clearing up. It became clear that even the Confederation of British Industry was not convinced by some of the reforms. The CBI has become more attuned over the last few years to *actual* skill shortages: to what they are and where they occur. The literacy and numeracy debates of the 1970s, which treated the issues in broad brush strokes while viewing skills very narrowly, have been made to look a little like those we used to have about Latin and Greek.[2] There is a greater appreciation that we need specific skills for specific jobs. Yet the reforms quickly veered away from improving the (expensive) technological competence of children when the ever popular chant of (cheap) 'long division' mobilized the masses.

Parents are quickly politicized. They are often referred to as a homogeneous group with the same educational ambitions for their children. Yet parents can be industrialists, milkmen, teachers and even politicians. Where the issue of relevance emerges a painstaking process of dialogue needs to be established concerning curriculum and community.

IMPACT either pitches itself into such dialogues or it precipitates them. In a project that seeks simply to involve parents in school-based curriculum work, discussions being at the points where they are most keenly felt and matter most. Parents and teachers find they need to ask questions and find a common parlance for discussing the curriculum and its issues. Schools can be useful centres of discussion. Anxieties about the shape of the future and constructing the appropriate curriculum for individual children should be met by considering not only where the child and the school is at present, but also what we want our lives, and our children's, to be like in the future. Industrial firms assess school leavers to fit them to jobs. When it becomes manifest that the form of assessment may match neither what the pupil has acquired nor what the job actually requires then questions need to be asked before we unthinkingly demand that schools change. A school is a useful place for such disclosures. Such anxieties and opinions are the natural points for creating a means for parents as outsiders to gain entry to curriculum discussion from their perspectives.

In their turn, schools need to look more closely at the lives of children, and adults, outside school. The conditions in which skills are acquired and deployed in everyday life are different at home and in the classroom and on the street. But as teachers we need to know what questions to put to parents, from a viewpoint within the curriculum, to enable them to give us sensible answers. Hopefully we can get beyond the curriculum clichés. The connection between 'practical maths' and the so-called 'everyday world' proves to be a poorly thought out one. That exotic place in the vocabulary of schools called 'the outside world' is often used to justify aspects of the curriculum. Involving that great outside in the curriculum will perhaps wobble some of the justifications — and if that happens we will just have to think again. And then again.

Let us now turn to the historical background of IMPACT and parental involvement.

Years of Social and Professional Change

It would be very convenient if we could sit and tell a neat story about how and why IMPACT started when it did, and what historical factors led up to it. It would be very convenient to state how educationally enlightened we had all become, how politically mature, and how schooling had become more socially outgoing. We could then present IMPACT as being perched on some pinnacle of progress in which popular opinion, social prudence, academic research and an avant-garde curriculum had all suggested the time was right

Introduction: Debate, History and Challenge

for involving parents in the ordinary processes of the curriculum to the advantage of children, schools and their communities.

Alas, there is no neat history of this kind. Sometimes there are overriding political reasons, not educational ones, for doing one thing rather than another. IMPACT emerged in 1985 at a time when greater political attention was being given to the relationship between parents and schools. Research into child development certainly added impetus to this political attention in debates on education. But arguments could equally be mounted to keep parents out of the curriculum. Indeed, one such argument we have heard a number of times is that research into child development has given us the style of mathematics curriculum that began to be evangelized in the early 1980s. It has been put to us that since parents are not familiar with it they will inevitably confuse their children if allowed to get too close.

Whether or not one takes issue with such statements on educational or academic grounds, the fact remains that assertions of this type appear to be naïve in view of the cultural, social and political changes of the latter half of the twentieth century.

Cultural Change

Since the 1950s there can be hardly any sphere of cultural activity which has not been affected by the challenges posed by those on the receiving end of the professional institutions: medical, religious, educational, psychiatric and so on.

Whether or not lasting fundamental change has occurred as a result of these challenges is beyond the scope of this book. Such changes that there are could be summed up in the words 'consultation' and 'accountability'. The necessity for consulting people in matters that will affect them and in being accountable for actions taken now appears to be part of the consciousness of the professions.

Authentic consultation and accountability cannot be guaranteed and many people have stories to tell about mistreatment at the hand of the professions where only lip-service had been paid to these sentiments of equality. However, grievances against one's fate at the hands of the professionals find ready expression in pressure groups and 'watch-dog' agencies for those that feel empowered to make a personal challenge. In education alone, 1954 saw the formation of the National Confederation of Parent-Teacher Associations, 1960 the Advisory Centre for Education, and 1962 the Campaign for the Advancement of State Education.

Unfortunately, the basis of our increased awareness and scrutiny of what the professions get up to tends to retain a confrontational character. It seems we struggle more to be heard or to gain satisfaction than we do simply to participate. In the early 1960s autistic children, who at that time were thought to be ineducable, were hospitalized where appropriate. Parents struggled to convince the authorities that these children should be in schools. Now many autistic children are in schools receiving special education. It seems that our desires for our children, as in this case, have to mean mobilizing our attentions in a political frame of reference.

Advances have bene made in the interpersonal spaces between, say, pregnant mothers and doctors, teachers and parents, and mental health workers and the mentally ill. It is more usual now for the patient or 'client' to be listened to during diagnosis and discussion of treatment. This phenomenon is far from universal and is practised unevenly where it does occur. We cannot say that clients receive fairer or better treatment, but we can say that people prefer to feel in some sense in control of their fates and not the passive victims of impersonal processes.

If anything, we might argue that everyday language now contains many more words, phrases and clichés with which people can express victimization or lack of control at the hands of the professions. This in itself is a powerful motivation for politicians since, after all, most victims, passive or otherwise, are still voters.

Educational Change

In themselves, consultation and accountability do not add up to the participation of 'outsiders' in processes that affect them or their children. Arguably an autistic child could be happier in a hospital unit than if 'made to work' in a school unit. An over-simplification of a painful dilemma for a parent we agree — but it suggests that a parent's involvement in their child's development has scope well beyond the shifting from one kind of institutional setting to another.

In mainstream education a desire to participate in the learning and development of children manifests itself in the phenomenon which many teachers feel uneasy about: parents buy reading and number books from which to 'teach' their children at home.

Teachers fear that this 'interference' can confuse the child. Parents may see maths, for example, simply as arithmetic and algebra — children can become confused about terminology and may get confused between ways taught at home and at school for addition and subtraction. Furthermore, many teachers see teaching as a process of enabling children to develop skills and strategies. Parents are often thought to view teaching as instilling information.

Topical though these issues may be, they occur now in a context which has developed amid widespread discussion and education innovation over the last twenty years.

Introduction: Debate, History and Challenge

So What is Parental Involvement?

Are parents clients, customers, partners, nuisances, unpaid helpers, insiders or outsiders? We argue that this cannot be answered in any simple way. On different occasions and in relation to different problems and issues that come up in a school any individual parent can occupy any of these roles.

There is always a tendency to try to simplify the situation because decision-making in education is such a complex business. Discussed more fully below, there is a strong tendency in the 1988 Education Reform Act to view the parent as a consumer of an educational service. In many ways this makes for more simplified and stream-lined decisions in some school matters such as presentation of the school prospectus or financial management.

But ERA was passed during a decade in which the role of the parent in education had been discussed with some passion. When the mist clears in future generations, we may be in a better position to see why politicians and professionals would like parents to occupy certain roles rather than others.

Political Change in the 1980s

We may summarize the development of education legislation once more in the word 'accountability'. At least since the Taylor committee reported in 1975,[4] it has been acknowledged that parents needed more uniform representation on school governing bodies. Nationally the picture was of very uneven representation. The Education Act of 1980 requires that parents are elected onto governing bodies. The act cannot force people to come out and vote however and typically this exercise has had tragi-comic consequences.

The Act also provided for the parents' right to choose a school for their child. The Act supplied weight to this provision by also providing for an appeals committee to whom disgruntled parents may refer if refused their first choice.

The Education Act of 1986 consolidated the parental rights of 1980 and set out compulsory forms of information that governing bodies had to supply to parents. The governing bodies now had the power to determine curriculum policy. To some extent these provisions have been overtaken by ERA which devotes a whole section to the information requirements of parents and other interested parties.

The ERA sets out a National Curriculum which has been described as a 'curriculum, assessment and reporting PACKAGE'.[5] Detailed assessments are made and reported to parents at specified intervals. Within this framework standard parental involvement will be as users of detailed 'assessment information' about their children. Assessment information will become the basis on which many decisions in education and discussions about children will be made. The National Curriculum is tied to a new system of financial management delegated to governing bodies within schools. A new complaints procedure is also set up on the model of those developed for consumers and traders. We should like to pose the question as to whether a consumer complaints procedure is the most useful framework for a pattern of parental involvement.

In summary, the educational reforms of the 1980s mean an increase in the control of the curriculum and the school by those traditionally seen as 'outsiders'. The governing bodies of schools are the 'focal media' of these changes in which increased control has meant increased accountability. In this context the nature of parental involvement can undergo a number of unpredictable transformations. So, for the moment let us back-track to more familiar territory.

What is meant by parental involvement at any moment in history is subject to widely varying interpretations. Perhaps in jest one may facetiously remark that parents supply the raw material for the education industry. Joke? Well, maybe, but we will return to this shortly. During the 1980s the contrast between styles of involvement had been identified as that between 'client' and 'partner'.

Parents as Clients

The parent as client is essentially an outsider who has to be 'dealt with' by the experts who are 'insiders' to the process of education (teachers, education administrators, advisers, psychologists, etc.). Involvement as a client means receiving information about school, about decisions already made and dealing at most with peripheral aspects of education — peripheral that is to the curriculum.

Parents as Partners

This concept achieved widespread approbation in the 1980s. It covers the more active roles a parent might occupy in fields such as school management or involvement in the curriculum through shared reading perhaps. There have been some schools, filled with the pioneer spirit, which have brought parents in to contribute to teaching and learning across the curriculum.

Other Sources of Influence

The Plowden Report of 1967,[6] advocated that parents should be welcomed in school and envisaged the practice of regular review meetings and the need for systematic and comprehensive reports on the progress of children. In the late 1970s research revealed that in 80 per cent of 1700 contacted schools 95 per cent of primary schools held parent evenings and open days with attendance levels of over 75 per cent.[7]

Introduction: Debate, History and Challenge

In 1978 the report of a survey carried out by HM Inspectors of Schools entitled *Primary Education in England* stated that,

> Parents helped teachers in nearly a third of the 7-year-old classes and in just under a fifth of 9- and 11-year-old classes.[8]

They go on to report that the incidence of parental help was lower in inner city areas than in rural areas. Parental involvement with learning '... took the form of assisting with practical subjects or hearing children read'.

The Thomas Report

In 1985 the report of the Committee on Primary Education *Improving Primary Schools* was published by the ILEA.[9] Norman Thomas had been Chief Inspector for Primary Education. The report is a candid and balanced account which devotes the fourth part of four to 'Parents, Children and Teachers'. The section on 'Agreement between teachers, parents and government' reads:

> No one can suppose that all parents and all teachers can see eye to eye all the time.... What is necessary is that children sense that their parents and their schools are in broad agreement about their education ... and are prepared to work together in its interest.... This calls for a degree of tolerance ... and a degree of compromise. Schools and parents — and the community as a whole — should be reasonably in step in their educational opinions and practices. (para 4.2)

The report acknowledges the manner in which information about pupils passes in informal contact. But it also discusses the issue of 'Parents as Collaborators'. Paragraph 4.25 says:

> ... the view has grown that parental involvement with schools should be increased to the point where parents share in the activities of the school both with respect to their own children and more generally. We are persuaded that this trend is right partly because there is convincing research evidence in favour going back over twenty years....

The section goes on to discuss 'Homework and primary school children', in which we discover that there is a desire among parents for homework at the primary age range. The committee recommends work of the 'PACT type'. Here parents are involved in their children's learning to read through a structured dialogue about books the children take home. The uneven quality of this work was also recognized by the committee. It warns that special steps would be needed to make sure the children of parents refusing to help would not lose out.

The report also suggests what any shared work might look like if the curriculum area was other than reading. We shall return to this after considering briefly parental involvement in reading.

(For further detailed information on parental involvement and associated issues up to 1983 see Sheila Wolfendale's book on *Parental Participation*.)[10]

Parental Involvement in Reading

The intimate involvement of the home with the life of the school suggests to most people the image of parent and child sitting together to read a book brought home from the school. A fairly simple thing to organize one might think. Not at all. All commentators and researchers on the subject note variations on the 'orthodox' ways in which schemes such as PACT operate.

The research findings on reading attainment in pupils who have been involved in PACT are well known. Improvements in both the reading ages and children's attitudes to reading were widely reported. Yet these findings still draw notes of caution from specialists. And as we have observed, practitioners are likely to point to the wide variation in practices.

The impressive achievement of PACT, research findings aside, is its contagious spread through the land. There are intelligent examinations of parental involvement in reading which discuss the various types such as 'paired-reading', 'shared-reading', etc. You are referred to Keith Topping and Sheila Wolfendale's *Parental Involvement in Children's Reading*.[11]

In a very friendly book, *Parent, Teacher, Child*,[12] Alex Griffiths and Dorothy Hamilton in 1984 set out the philosophy and guidelines for PACT. They take a view of parents which the IMPACT team has discovered to have a surprising welcome among teachers. In the first chapter they say,

> Children learn first and foremost from their parents. In this respect all parents are teachers — and very effective teachers they are. Arguably, children learn more from their parents in the first five years of life than they do from their schools in the next ten.

Potentially explosive stuff you might think. There is a strong sense in their book that parental involvement in the education of their children is desirable in itself, everything else aside. A view which we would endorse. The notion of parents as teachers is more complex. The research of Jack Tizard and Jenny Hewison[13] in 1982 suggested that what parents *actually* did that appeared to improve attainment was to 'listen' to their children read.

So far our ramble through the historical background to IMPACT has indicated that politics, concepts of parents, and developing curricula have each had a hand in fashioning what parental involvement in education means. Griffiths and Hamilton's statements about parents as teachers (and their implication that listening is a form of teaching) indicate we must at some stage consider the way in which thinking about child development and learning processes contribute to the story. This we attend to in chapter 1. For the moment we feel it is time to focus on the beginnings of IMPACT.

Introduction: Debate, History and Challenge

A Brief History of IMPACT

So legend has it, Hampstead Heath was the place where Ruth Merttens, contemplating the Thomas report of 1985 and the PACT experience, first began to formulate what a mathematical version of PACT might look like. In a large footnote, legend also relates that Jeff Vass had been attempting to grasp how 7-year-olds communicate with peers while solving problems, and how teachers communicate with primary-aged non-communicating children.

The Polytechnic of North London became the venue in which Ruth began to formulate IMPACT as a pilot project. She was clear in one aim: that IMPACT was about involving parents in the curriculum in a structured and central way using mathematical activities as the focus of the attention of, and communication between, parent, teacher and child.

The Pilot Year 1985–86

In September 1985 there were six inner London schools doing IMPACT. There were ten by the time we left inner London at the end of the summer term 1986. The pilot year allowed us to work out the shape of IMPACT in collaboration with the staff, parents and children of these ten enthusiastic and supportive schools.

Materials for working at school and at home. On PACT, children take home books. For maths there have to be materials. But not just any materials will do. We discovered early on that activities children do at home with their parents have to be adapted for the home context and preferably designed with the home specifically in mind.

The Thomas report (para. 4.29) stated that the type of work that 'can be done with advantage is the extension and follow up of work begun in the classroom.'

The outcome of our findings with materials to use at home is more complex in that several factors appear to be involved:
- style of teaching,
- form of preparation,
- nature of task (e.g., open-ended, investigative, problem-solving),
- task theme (curriculum-oriented, topic-oriented, home-oriented),
- form of follow-up work.

During the pilot phase we recognized the need for in-service training for both designing materials and the mathematics curriculum in general. There was, and is, a great need for primary maths INSET. We came to the view that IMPACT was a good medium for supplying the sort of INSET that was required not just to do IMPACT but for what was construed as 'good primary practice' in general.

We discovered that the design of tasks appropriate for use at home meant that all of us were concentrating on areas of the maths curriculum which had been specially identified in the HMI Mathematics 5–16 document.[14] Such features as problem-solving and transfer of skills particularly interested us; and in any case these features always figure large in discussions on the 'relevance' of the curriculum.

Transferring skills from the context in which they are first presented to another is identified as a primary objective of education in general. Everyone desires this. However, homework in the traditional sense often means utilizing skills acquired in school to practise further problems of the same kind as if one were still at school. This cannot be said to be an authentic instance of skill transfer. On the other hand, imagine that a child has learnt to perform recognitions of 'bigger than' and 'smaller than' while using different sized blocks at school. If s/he now goes home with an activity asking her or him, together with a parent, to find three objects smaller, and three objects bigger than her or his shoe we are more likely to provide the conditions in which transfer of skills to another, different, context might occur.

Parents and children. The families that got involved during the pilot year responded exceedingly well to what was being asked of them. Having set up regular 'follow-up' parent meetings in each of the schools, we attended as many as we could. In those we did not attend, we asked teachers to keep records of what points were raised.

There were 300 families involved altogether. They belonged to a wide range of ethnic backgrounds. There were single-parent families, and both working- and middle-class families.

We could discern no differences in response rate between classes or ethnic backgrounds. The response rate (calculated as the number of children performing a task at home each week) averaged out to 80 per cent over the course of the year. Admittedly, the IMPACT team was in the schools a great deal, and the project was kept in high profile throughout the year. There were dips in response rate too, below 40 per cent in odd weeks, for which we blamed inappropriate tasks, or inadequate preparation in the classroom. Readers may recall that 1985–86 was the year of teacher action.

The children who regularly did not participate consisted of precisely those who did not do PACT, as well as two children whose parents objected to the philosophy of a project which required parents to participate in the educative process (one of these parents was a teacher).

One teacher, writing in an article published during the pilot year, wrote:

> I now feel more confident about mathematics, and this clearly affects the children's attitudes.... Our school has a strong tradition of parental involvement, but mathematics is the one curriculum area in which communication with parents is important yet most difficult.... There is more understanding on the part of ... parents of their own child's learning patterns and there is a focal point from which to talk to me about specific interests or problems.[15]

Interestingly, this teacher also writes in the same article about some parents who, on occasion, have felt cheated by some of the activities because they had not received the 'pages of exercises which they had expected'. The

Introduction: Debate, History and Challenge

contrast between the expectations of teachers and parents began to emerge in stimulating, if at times uncomfortable, ways.

Parent meetings proved to be fascinating events in which some of the fundamental questions about the nature and purpose of education were discussed. They also provided the arena in which new methods of teaching and new styles of curriculum could be debated.

Children as tutors. The children claimed to enjoy most of the activities. But the most interesting thing lay in the discovery through response sheets and parent meetings, that in a substantial number of tasks children were acting as the tutors in the home. It turned out that the children perceived themselves as 'knowing more' about what had to be done than their parents. Even nursery-aged children got into the role of initiator and organizer of the activity in the home. We carried out some video-based observations at this time (reported elsewhere) both in homes and in schools. Analysis of the video tapes was not necessary to tell us that children were operating in fundamentally different ways at home than in school. Despite the freakishness of video cameras and IMPACT personnel discreetly placed in living rooms, we observed qualitative shifts in the children's approach to and sense of mastery over a task which had been introduced to them during the week at school. We will discuss this again in chapter 2.

The IMPACT Project: 1986 to 1989

If you have ever been to one of those parties where someone always seems to be walking about with a dust-pan and brush bothering everyone else you will be able to sympathize with our experience of IMPACT in the pilot year. On a few minimal premises a great deal of innovatory practice and research data had been generated. The pilot year had seen the development and trialling of new materials, forms of parental feedback, forms of classroom management, and forms of peer and child–adult interaction.

We had not the time to initiate and follow-up developments in schools, collect data and hold down our jobs! In effect, we 'bothered' the developments as they were becoming established — the worst of both worlds if you consider the hang-over as well.

So, part of the pilot year was spent marking out the financial and logistical groundwork of a 3-year project. We wanted the 'IMPACT Project' to be primarily an interventionist project with a large research and evaluation component.

An Intervention Project

In order to develop materials with teachers in the classroom for the home a certain amount of primary maths INSET had to be provided; hence a substantial proportion of the financial base of the project came from supplying maths INSET in the three LEAs which had been first to respond to an opportunity to take the project. The LEAs were Barnet, Oxfordshire and Redbridge.

We worked in close cooperation with the maths advisory teams in each LEA and in the case of Oxfordshire with the Home-School Liaison advisers. Each LEA had its own 'development' plan in that each had different ideas about where INSET would be used, how schools at this stage would be chosen and how the IMPACT initiative would develop throughout the authority. Each LEA expressed strong commitments to the principles of parental involvement and to extending maths INSET.

If there were differences between authorities there were greater differences between schools within each authority. We have always stressed that each school should have a 'custom-made' version of IMPACT to fit with the curriculum of that school. Before the National Curriculum all schools had very different approaches to mathematics teaching in terms of both content and method. We stressed that IMPACT should be 'Puddleglum-IMPACT' or 'Stoneyvale-IMPACT' or 'Orbiston Parva Primary-IMPACT'.

However, we stipulated that the following features had to be present in order that what any school did could be called 'IMPACT'. There should be:
a) PLANNING — each IMPACT teacher was asked to plan the maths curriculum at least a few weeks ahead.
b) INTRODUCTORY PARENT CONSULTATION — IMPACT had to be introduced to *all* the parents whose children would be taking part.
c) PREPARATION — any activity to be done at home should have some connection with current on-going classwork, and the children had to be explicitly introduced to some aspect of the activity.
d) ACTIVITY VARIETY — there should not be just maths 'games' sent home which practise skills even if appropriate on some occasions. There had to be investigative and problem-solving work and data-gathering for classwork.
e) HOME FEED-BACK — opportunity had to exist for both parent and child to report back on how the activity went and what they thought about it (*see* research section below).
f) FOLLOW-UP — classwork following a home activity had to incorporate and develop what had been done at home.
g) PARENT MEETINGS — at least twice-yearly parent meetings to discuss progress should be held.
h) TUTORING BY CHILD — the bulk of IMPACT work at primary level had to be designed in preparation and in home activity so that the child became initiator and tutor in the home.
i) REGULARITY — sending activities haphazardly demotivates both children and parents and lowers the responsiveness of the home which in turn disappoints teachers.
Schools had great freedom to work out their own approaches to each of these features of IMPACT.

Introduction: Debate, History and Challenge

A Research Project

We organized our timetable and the day-to-day operation of IMPACT in the schools together with a continuous 'Evaluation, Monitoring and Research' (EMR) procedure. Briefly, we wanted data to be collected on each aspect of the project on a routine basis. A vast data-pool quickly accumulated, consisting of layered 'tiers' of data.

Tier 1: Weekly Data

For each activity sent home there was one questionnaire for teachers detailing aspects of preparation and follow-up. There were also forms for both children and parents to fill in concerning various aspects of the activity at home. These latter forms were designed to fulfil both research and feedback functions as stated above under necessary features of IMPACT.

Tier 2: Classroom Data

Details of teaching approach and INSET needs were kept by the LEA IMPACT co-ordinators. Other data for this tier include semi-structured interviews with IMPACT teachers and parents, as well as details of discussions at parent meetings.

Tier 3: Interaction Data

What children actually do in peer and child-adult settings in school and at home has been recorded on video. We confined this activity to a small number of selected schools.

The purpose of EMR was to provide a base from which to begin an evaluation of the project both for ourselves and for the schools and LEAs to use. But it was also there to conduct research into the fields of cooperative learning and peer tutoring which lie at the heart of the learning and developmental processes of IMPACT.

We have been fascinated by what has come out of the project. The full outcomes of our research activities are being published over the next few years. In the meantime we offer the following by way of illumination of some of the themes treated so far in this book.

Some Findings

In 1987 we started with six schools in each LEA. In Oxfordshire there are small village schools with a total of 90 children as well as large JMIs. Two classes from a large Redbridge primary gave us 70 children. Expansion in each LEA meant in some cases taking on extra classes in a school, in others taking on extra schools.

The following figures are taken from our data on parent response rates for the year 1987–88.

Sharing Maths Cultures

Table 1.1 Mean Response Rate

77.5%

No. of children = 1804

This figure indicates how many children and parents responded by working at home on activities sent home weekly by the school. However, on many occasions a school would not send activities out in given weeks because they, for example, may have been involved in extended follow-up work. Some schools sent out work on weeks that other schools had decided to keep free of IMPACT work (half-term, Christmas, etc.). An approximation based on teacher estimates of how many children do the work at home each week brings the overall average figure to 86.6 per cent.

In any week the response rate can fluctuate between 40 per cent and 100 per cent. It is quite usual for many infant classes to maintain a steady 100 per cent response rate over a seven-week period. Dips can occur in this achievement for a number of reasons like family holidays, uninteresting activities and so on (bear in mind we were still at this time developing a bank of suitable materials and many teachers were relatively inexperienced in designing materials for use outside the classroom).

It is less usual for classes to operate at several consecutive weeks at 40 per cent, though this did occur in one school. Having said that, a number of factors contributed to this, especially difficulty in organizing follow-up work. Forty per cent to 50 per cent, it should be remembered, is still a very good figure for sustained parental involvement.

We found that there were consistencies in the response rate attributable to factors related to age. For example the overall average of 77.5 per cent can be analysed according to Nursery, Infant and Junior age ranges.

Table 1.2 Parental Response According to Age Group

Age Group	Mean Response %
Nursery	78
Infant	85.2
Junior	69.4

N = 1804 (33 weeks per school year)

The consistently lower figure for junior classes indicated to us that parents of children of this age-group perceived school work at home as essentially a solitary activity in the normal homework mode. In addition, designing interactive materials for this age-group is a great challenge given the nature of the junior maths curriculum. In the next few years nursery and infant

Introduction: Debate, History and Challenge

children who started IMPACT in 1987 will be coming through the junior age range. It will be interesting to see if patterns of home–school responsiveness can be maintained by schools with established links.

Accounting for Variability

In any school, as we have stated, responsiveness will fluctuate for a number of reasons. We have mentioned that the type of activity influences whether child and parent will do the activity at home. Some tasks will make less sense to parents than others because the rationale for them is to be found in what the class are currently doing as their topic. If you have ever been stopped in the street by a market researcher and asked a series of apparently disconnected questions you may well have become baffled and bored. The questionnaire *will* make sense to the people who designed it and you can be sure your responses will be eagerly inspected by the commissioners of the survey.

Some IMPACT work can be a bit like this. We found that many parents found a short rationale on each activity helpful and motivating. Nevertheless, as responsiveness fluctuated we noticed other regularities in the classwork which seemed to go hand in hand with changes in response rates.

Within School Variability

If there were noticeable changes in response rates within any class that could not be attributed to poor activities we had to look again. Over the course of the year a pattern emerged which was paralleled in nearly all the schools.

For a number of different reasons there were occasions where teachers had been unable to integrate IMPACT into the on-going classwork. The IMPACT team had initially laid stress on making IMPACT a weekly or fortnightly event. The pilot year taught us that regularity is needed to sustain a project of this kind where changes in educational habits are desirable. To maintain momentum teachers sent out activities each week. Most of these activities arose from on-going classwork and were fed back in. Some were what we called 'holding activities': these keep the rhythm of involvement going but do not substantially affect the nature of class follow-up work.

Any activity becomes a holding activity, and not necessarily a successful one, when it is a supplement rather than a complement. Ideally, any task done at home is intimately related to classwork. And as we discovered, a lack of integration correlated with times of low responsiveness from the home.

There are data concerning extreme variability in a class's experience of IMPACT. In these cases where the response rate would pitch down to 10 per cent then up to 40 per cent then up to 70 per cent and then back to 10

17

Sharing Maths Cultures

per cent, we were dealing with classes of children who had English as a second language *and* whose families were in short-term housing.

In the graph (Figure 1.1) you will see the response rates in five schools plotted over the course of the year.

Subsequent Years

We were in a much better position to plan and cater for the next two years on the project which included methods of organizing parent meetings, editing materials, conducting INSET, and responding to the very different needs of the schools, given the experience and research we were able to undertake.

After IMPACT had survived the year of teacher action in the pilot year, the Project found itself facing the onset of the National Curriculum and the other reforms of the ERA. In many ways the preferred method of working in IMPACT pre-empted many of the stipulations of ERA such as the construction of teacher-plans and records which were already part of IMPACT practice. Naturally, information for parents was already a strong feature of our activities. The relation between IMPACT and the National Curriculum is discussed more fully in chapter 5.

Much of the advice and many of the techniques we have to offer are included in this book. But at the end of IMPACT the Project, we set up IMPACT the Network.

The IMPACT Network

Many more LEAs, schools and teachers wanted to operate IMPACT-like schemes in their schools than the number with whom we could be personally involved. Seeing the need for a nationwide network for the pooling of experiences, data, materials, INSET, etc. we set up, in September 1989, the IMPACT Network. Currently there are eighteen LEAs affiliated, as well as individual schools. Further details of the Network are given under the Resources Section in the Appendix at the end of this book.

The Future

Whatever happens in the future it looks like parental involvement in one form or another is going to be a strong feature of the educational landscape. The manner or style of that involvement is up to us. IMPACT schools have developed patterns of parental involvement which benefit not only the children but the schools and their communities. Schools find that they can work well alongside parents within the context of the curriculum. During a time of immense change in the shape of the curriculum we should welcome a strong sense of togetherness as communities.

Introduction: Debate, History and Challenge

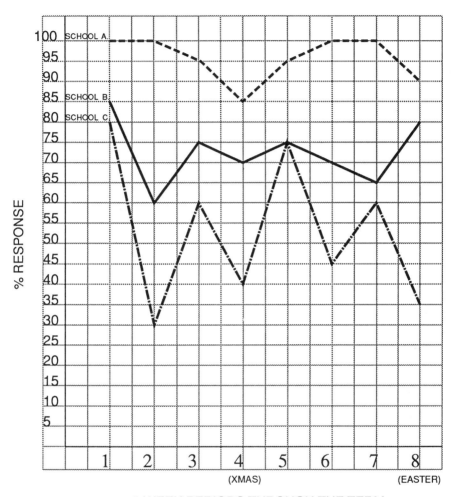

3 WEEK PERIODS THROUGH THE TERM

SCHOOL A.   SCHOOL C.
SCHOOL B.

The findings we have quoted here suggest that more than goodwill, there is a positive desire in parents to contribute very directly to their children's education in ways that they perceive to be relevant and important. IMPACT is about giving them and their schools an opportunity to do just that.

Chapter 2

Children as Tutors

> *Do we hear any worse jumble in the gabblings of fish-wives than in the public disputations of these professional logicians?*
>
> *Michel de Montaigne*

What Is IMPACT?

Ask any of the thousands of IMPACT parents, teachers and children, 'What is this IMPACT?' and the chances are you will get different answers from each of them! Naturally, there will be common elements in these answers. Many will see IMPACT as an activity or task which a child does at home with the help of a parent, brother, aunt or co-opted passer-by.

Ask either of the two authors of this book what IMPACT is and you may very well get different answers again. It is a bit like asking 'What is theatre?' One can point to a building, to a group of people, to a text, or one can refer to the role of theatres within culture in general. The same is true, surely, of parenting. What a parent is can be answered in many different ways. The same is true of shared reading even if ordinarily we tend to focus in on what the child does at home with the parent.

The problem with saying that IMPACT is 'a parental involvement project' or 'a trendy maths project' or 'a primary school homework scheme' is that it is all of these things and none of them. We are not being deliberately evasive on this point, just as honest as we can.

Indeed, we can point at classroom situations, at the process of doing tasks at home or even at sheets of paper and say *that* is IMPACT. Well ... yes ... but ..., it all depends not on the *what* but on the *who*. IMPACT will suggest certain strategies so that ways in which people can come together usefully, beneficially and, we hope, convivially, may be organized. Whatever methods or techniques we prescribe, IMPACT depends on who uses them and what they use them for.

In chapter 1 we quoted a teacher whose own attitudes to the teaching of maths had changed while doing IMPACT, who had noticed changes in the attitudes of the children and who had started to document the somewhat painful process of attitude change towards 'school-maths' in parents. Given a school's, or a teacher's, particular set of priorities and having read the literature on IMPACT, it might be decided to use IMPACT primarily to

Sharing Maths Cultures

change attitudes among staff, parents or children. Alternatively, a school may decide to 'do IMPACT' because another school up the road is doing it. In the first case IMPACT is then an 'attitude-changing project', in the second a 'keep up with the Jones's project'.

We can write down the how-you-do-it in this book. But what you get is an instrument. How and why that instrument is to be used depends on you, the agents. If anything, we would like to call IMPACT that process in which different groups, parents, teachers and children, come together around a mutual need — the education of children. They discuss the form, layout and strategic use of school, home and interactional environments.

IMPACT: A Skeleton Outline

If school life could be set to music every school would be a different symphony, each child, teacher and parent a different melody. It may not turn out sounding much like Beethoven or Mozart — but you could identify tunes and rhythms. IMPACT seeks to add its own particular rhythm to the symphony.

In Chapter 1 we listed the 'essential' features that a school doing IMPACT would need to implement. The list is as follows:
- Planning
- Introductory Parent Consultation
- Preparation
- Activity Variety
- Home Feed-back
- Follow-up
- Parent Meetings
- Tutoring by Child
- Regularity

The particular way in which each of these features is worked into school life, and the precise style each feature subsequently has, will depend on the character of the school.

Having said that, you will notice that some things need to be done before others. In addition, we would like to draw your attention to the particular rhythm that IMPACT adds. Below we make a sketch in outline; fuller details come in chapter 3.

Two Rhythms: Slow-beat and Fast-beat

Slow-beat. Over the course of the year teachers are involved in planning the curriculum and reviewing progress with parents. In practice this will mean planning at half-termly, or termly, intervals, then doing the classwork using IMPACT materials (either selected from a bank or designed by the teacher). Periodically, there will be parent meetings (two or three per year). At these meetings progress is discussed reflectively. Decisions will then be made about the shape of the next planning exercise and term's work. So, broadly speaking, the 'slow-beat' involves a yearly rhythm:

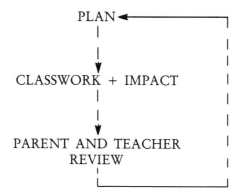

Fast-beat. Within the yearly cycle described above is the short cycle which lasts from two weeks to four weeks. Essentially, this cycle is concerned with work done in class, work done at home and work followed up in class. In any week's classwork there will be some preparation for a task that the children will carry out at home with parents. Once an activity has been selected or designed for that purpose it goes home with the child. The task is directly related to the work children are concurrently doing in class. The following week the work done at home is substantially incorporated into the classwork and developed by the teacher.

We call this process of 'preparation, home-task, follow-up' an *activity cycle*. Included in each activity cycle is some means by which parents can report back to teachers their experiences of the work done at home. Weekly informal conversations are excellent, given the opportunity, but standard practice on the IMPACT project involves written comments.

Activity cycles are adaptable to most styles of teaching and curriculum practice. They do not presume that teachers use an integrated curriculum, maths topics or schemes. Teachers using any of these methods routinely in the classroom can implement IMPACT activity cycles.

This then is IMPACT in outline. Chapters 3, 4 and 5 look in greater detail at each stage of the process outlined here and use examples of practice which have occurred in the Project.

Summarizing so far ...

Even though some essential features need to be present for IMPACT to work well and be of all-round benefit, these features fit in with and adapt themselves to the life of the school. Common-sense would suggest periodic consultations with parents — and some schools already practise this prior to

IMPACT. However, other features such as utilizing the children 'as tutors' to other children and to their parents during the activity cycle may require further justification. So, before we go on to consider activity cycles in greater detail we take a brief excursion into the world of cooperative learning and peer tutoring.

Children as Cooperative Tutors

From what has been said so far it will be clear that IMPACT is not primarily about mathematics materials. Mathematical tasks will, however, figure prominently in the rest of the book. Children always take home a task to do which will be set out in some way on a sheet of paper. Remember that the task to be done at home arises initially from something the teacher is doing in the class. Not everything that the class has been doing can be recorded in detail on a sheet of paper. Essentially then, the home-task is a narrower part of something of which the children will have had wider experience.

The onus is on the child to make available to themselves and their parents, aspects of this wider experience in the execution of whatever the actual home-task happens to be. The centre of what IMPACT is all about concerns the communication the children and their parents enter into while sorting out, managing and executing these tasks at home. We often say that the sheet on which the activity is written is a 'back-up' or *aide-mémoire* for the child and a communication from the teacher to the parent. The child is the central agent in getting the task organized and supplying its meaning as far as he or she is able. Theoretically, a good IMPACT task can go home on a blank sheet!

Talk Solves Problems?

Teachers today value more highly the opportunities that classrooms can give for children to engage in the sort of conversations alluded to above. However, it seems to be common experience that we recognize how useful collaborative and conversational approaches to maths can be when we are sitting at home reading a book about the subject. Yet we may feel slightly less easy in the middle of the busy classroom about how and where actual maths learning is taking place. Somehow we know it is good that children have mathematical conversations with each other, but we may not be too clear about what, exactly, is meant to be going on. For one thing, maths often means using clearly defined procedures and it puzzles us that such things can emerge in the context of the apparently messy conversations that, say, 6-year-olds appear to have when sitting around a table doing number investigations.

Language and maths. A sense of the importance of language in mathematical work is becoming more widespread. It is difficult to see where mathematical training could occur outside talking and reading about it. Essentially, we recognize that we use language to organize our 'doing of maths' and to understand the significance of what we have made of the practical or abstract work that we have done. It does not seem to be true that first we work out, in language, what we have to do mathematically, then do some exotic computation and only then put back into language what we have just done.

It seems language has a much more intimate connection with all stages of mathematical work than one might expect.

The Social Context of Maths

If we follow up further the relationship between maths and language, we arrived at a point where we want to say that in order to perform most tasks in life we have to make sense of them. Language then becomes important because it is the tool we use most often to make sense of the situations in which we find ourselves. Language is a social tool. Its first job is to help us make sense of each other! This definitely does *not* mean, however, that there are ready-made 'nuts' in the world around us, social or otherwise, for linguistic 'spanners' to get to work on. Yet language helps to coordinate the activities that people find themselves coming together to do. 'You wash, I'll dry' is a familiar piece of coordinating language you might overhear in somebody's kitchen. If, from personal experience, you find that example unrealistic you might consider this: 'I've just washed up, what about you putting the kids to bed?' The latter not only coordinates (or tries to!) the activities of another person, it attempts to provide a meaningful justification for the second person to enter the domestic agenda!

Asking a child if he or she thinks there will be enough time for a story after the washing-up has been done *and* after he or she has put away his or her toys but before mum has got off the sofa to put the kids to bed, clearly relies on the child knowing a great deal of social (in this case family) routines and conventions. These conventions form part of the substance of the problem the child has been asked to solve. But they also serve to locate the more abstract structure of the problem within the familiar and meaningful. As Margaret Donaldson[16] and others have argued, it is in circumstances like these that children solve problems which they might otherwise find difficult.

Notice though that it is not the 'concreteness' or the 'abstractness' of the problem which the child finds easy or difficult respectively. It is the child's familiarity with, and current proximity to, the particular conventions that form the background to the problem that are said to make things either easier or more difficult.

Because maths is embedded in language and other social conventions, the latter are able to lend maths their peculiar qualities, among which is flexibility. Flexibility in language may give us flexibility in approaching a problem. At least that's the promise. In practice a limited command of language may narrow the choice of strategy we take to a particular task. Social conventions can also be stifling when trying to solve problems. It is often said that, typically, skills acquired in school are not easily transferred to other situations. It is also said that school-maths is viewed too often by children as something that is only useful and meaningful in school.

Children with learning difficulties suffer particularly from not being able to cope with what are construed as contextual changes — having learnt in one context, a similar problem given in another context can, all too easily, produce misrecognition and confusion. Ironically, it may be the case that our techniques for uncovering where the problem actually lies tend to direct us always to the child and to what is inside his or her head rather than to the very different contexts he or she inhabits. For further discussion of this see Merttens and Vass.[17]

So far, in discussing the social context we have talked about language and social conventions but not much about people. It is when people come together around a particular problem that talk becomes almost irrepressible. But the type of talk we use does depend, to some extent, on where we happen to be. There is one kind of vocabulary we use in the kitchen which is rarely used in the night-club; we have methods of talking in the staff-room which would be out of place in the bedroom; we phrase things one way for car mechanics and in another way for lawyers. Our competence in any of these situations can be highly variable. You may have met the chap who can render Latin poetry into any one of six modern European languages yet who is disastrous when attempting to describe to a plumber over the 'phone what is wrong with his cistern.

In the last thirty years we have come to a realization that the social context in which we attempt to get things done will influence or determine the way in which we will actually do things. You may be familiar with those studies (Donaldson, 1978)[18, 19] in which children have been presented with a problem that they have been unable to solve in one context but have been more successful in another. Take a long, narrow cylinder and pour a pint of water into it in front of a 4-year-old. Then pour the water into a short fat container and ask the child 'Is there more, less or the same amount of water as before?' Typically, the child gets the answer wrong, demonstrating an inability to conserve. On the other hand, if a plausible reason is introduced to account for the pouring of the water from one vessel to another very different results are reported. For example, if the adult says to the child 'Oh look, the tall narrow jar has a crack in it . . . better pour the water into another jar' and then asks the same question about whether there is now more, less or the same amount of water in the uncracked fat jar the chances are more in favour of the child saying 'the same' (Hundeide, 1985).[20]

In these situations the child is dependent on what the adult does and says as well as on observing what are thought to be the mere physical events on the table. The child's use of what the adult does or says turns out to be extremely important in how he or she responds.

Adults Instruct Children

Since the 1960s attention has been focused on how different kinds of language in different situations have affected the way adults and children manage problem-solving and a variety of other tasks. One kind of situation that we apparently carry around with us is our social class. In the 1970s Basil Bernstein[21] demonstrated that what we define as working-class and middle-class can each be correlated with different patterns of language use. It is not suggested that one pattern is superior to the other in potential to solve a problem — but typically each will use different techniques or strategies.

Hess and Shipman, in 1965,[22] observed mothers instructing their children in a number of tasks involving moving blocks to create required designs. In this research a pattern emerged in which working-class mothers typically coerced or controlled the conduct of the children in a didactic way without giving reasons to the child for any instruction given at any time. Middle-class mothers tended to be more motivating, offering explanations for the instructions given to the child.

For a parent and child from either class, of course, the task would get done. But the question remains about the benefit to the child and the extent to which exposure to instruction of either type leaves her or him with the facility to cope with similar problems on their own in the future.

It will not be doing justice, but we can summarize what we need from the research quoted above if we say that the contexts in which we normally live, to some extent, provide us with habits of language. These habits, in turn, provide us with different facilities and techniques for solving problems. We normally exploit our habits in the contexts in which we normally use them. Some habits can be used in a variety of contexts.

New Ideas in Child Development

Piaget, though still widely discussed in educationist circles, tends to be given shorter shrift these days. Piaget is read as though he were making universal claims about the way in which children have to develop. The child development literature is now overflowing with studies showing how, if situations are organized in amenable ways, children can do things ahead of the time one might have predicted from Piaget's developmental 'time-table'.

To some extent the great Russian psychologist, L. S. Vygotsky, who was writing during the first half of the twentieth century, has come to occupy the seat vacated by Piaget as guru of child development studies. Such fame has, if Piaget is anything to go by, unfortuante side-effects on how one's work is read and developed (Sutton, 1983).[23]

Vygotsky is seen as prioritizing the organization of social relations among people (in whole communities) within discussions of intellectual development.[24] This is a difficult notion. To over-simplify, we have to try to imagine that any activity we can organize and perform on our own is a competence we acquired firstly in a social setting. When we get into a car as driver for the first time we are entirely dependent on having what we do 'organized' through the speech of our driving instructor. He or she will give us a series of spoken instructions. Eventually we can run through the sequence on our own. Finally, when driving becomes second nature and we have made it 'our own', the nature of the original sequence can change. We find ourselves looking in the mirror *as* we reach for the gear-stick *as* we start to depress the clutch *as* we begin to move our foot from the accelerator. Skills acquired in many situations, from tennis to brain surgery could serve as examples.

In those examples, the social setting was a spoken one in which two adults allowed the organization of the activity, driving, within speech. Adults and young children also use speech, but very often there will be a greater practical sharing of the activity under way. Children can acquire competences from these less well-defined circumstances. Less 'defined' is used in the sense that the control of the situation is not dependent on all instructions being in speech; and it is not always necessary that the child knows precisely what the activity is meant to achieve. In teaching, great stress is often laid on a child being first able to understand the concept behind a process before they use it. The authors of this book find the word 'concept' misleading when it comes to the discussion of what children understand.

Setting, Language and Maths

Returning to our main theme we can now have another look at the situation in which children acquire mathematical skills. Let's try to imagine the busy settings in which most competences we acquire are actually acquired. Any activity a child is asked to do will use some skills s/he already has and may require others s/he does not yet have. In a classroom, or in the home, this will mean that the child will be operating on several 'fronts' at once. The child will be:
 i)   coordinating the material resources of the setting;
 ii) coordinating her/his actions with an adult or other children;
 iii) using language for a variety of purposes, including (i) and (ii);
 iv) using more precise mathematical skills such as adding and subtracting.
All this means a great deal of 'management'. If we focus more closely on maths we can pull out the following features as always being relevant to the on-going development of a child's mathematics within the curriculum:
- Using *mathematical techniques* (adding, subtracting, etc.).
- *Choosing strategies* and *techniques*.
- Using *private language* to coordinate those techniques.
- Engaging in *dialogue* with adults and peers.
- Understanding how one's current activity fits in with or belongs to the *social setting*.

Each of these feature at different strengths while a child is doing mathematics.

A child can pass back and forth from one to another in the course of, say, solving a problem. Initially, a child may look at a problem and launch straight into choosing some technique s/he thinks may be appropriate. Stopping short, s/he may then start to discuss with peers what they are doing. Finally, s/he may focus her/his attention on the social setting to seek clues to the meaning of the activity.

In group work the movement through each of these features in all directions among the children makes the actual course or progression of a group activity difficult to trace for a non-participant observer. Yet there are special characteristics that belong to situations in which two or more children or adults work together that we should now discuss further.

Cooperative Learning

We have already talked about studies in which mothers have assisted their children to solve problems. We noted that mothers from different communities will bring different linguistic resources to bear in these situations and that the problems will be solved in different ways.

Effectively, that sort of situation was one in which a parent, having a general grasp of the problem, tutored and cooperated with their child to achieve a solution. We will discuss adult and peer-tutoring below. For the moment we will concentrate on what happens when children come together to solve problems cooperatively.

In our own observations of children collaborating on activities we would say that surprising things happen. We once watched a group of five 6-year-olds learning a game new to them in which the object of the game was to lose the amount of money they had been given at the start. A working knowledge of totals is required throughout and also the ability to engage in strategic planning.

While the game was being learned by the group, one child set himself up as organizer of the materials and the other children. Now occupied, he was unfree to manage other aspects of the game. Two other children passed the running totals between themselves, relinquishing responsibility for the running total in order to try out a new strategy. Meanwhile the entire group was attempting to negotiate what the object of the game actually was, and what 'winning' looked like. As new theories about the purpose of the game were generated by the children the roles described above would be swapped. To further complicate the issue we noticed that several different versions of the game would be attempted simultaneously! The children cooperatively acquired the knowledge to play the game to the extent that each of them was able to teach it to another group of children in its *entirety* irrespective of the actual roles that they themselves had most experience of while learning the game. For example, one of the children who had looked as if he had participated least in the game, only contributing the odd calculation during the group conversation, was able to teach all aspects of the game to another group, including one or two useful strategies for winning the game.

Several things about this phenomenon surprised us. We normally presume that problems are solved in a fairly linear fashion. Any piece of maths can be broken down into a set of sub-skills: subtracting 87 from 92 may require recognizing and using number bonds. We are nervous about 'lifting out' particular parts of a problem-solving sequence from the set sequence in which we feel we need to teach subtraction. But when we observed these children they *played* with the linear arrangement of the problem-solving sequences.

We have not pursued our observations, although others have (*see* Wood, 1978),[25] but adults can become embarrassed when asked about the actual strategies they have used to solve a problem. If we were to ask you to state the relationship between A and B if A were B's mother's brother's sister's cousin's great aunt, you might go about it in a number of ways. We have the idea that there is a 'proper' way that requires algebra. People feel they have cheated if they solve it by mapping A and B onto their own family tree, for example. This is an extraordinary phenomenon which should alert us to our prejudices which constrain us to probing problems in narrow and stultifying ways.

Cooperation, Competition and Group-work

The game we discuss above is itself essentially competitive. The banner of cooperative learning does not need to imply that the situations in which children learn are necessarily cosy or friendly! Cooperative learning can, and does, take place in situations where the nature of the task requires children to compete. Furthermore cooperative learning is not a sufficient justification for group work as a dominant mode of classroom organization. if we take Vygotsky seriously then we have to consider that *one* child, when exercising an acquired competence alone, is still engaged in what is essentially a social activity. He or she still uses private dialogue, may still play with elements of the problem and so on. As teachers who employ group-work methods, on the one hand we need to remain sensitive to the relationship between the child's acquired and developing competences; on the other, we must bear in mind the variety of communicational possibilities and variations to which any task may give rise.

The importance for us of relating types of tasks to children's acquired and developing competences involves us deeper in the nature of the processes which occur in cooperative learning. Specifically, the relationship that evolves between two or more people when engaged on grappling with a new task or problem is supremely important for the process of learning. To this relationship we now turn.

Cooperation, Learning and the 'Transfer' of Knowledge

We have already given some indication that children develop competences while learning cooperatively. Vygotsky, and many developmentalists (*see* Wertsch, 1985)[26] feel we should distinguish between what a child has already acquired (demonstrable as what a child does with a problem alone) and how s/he will perform while operating with a more capable peer or adult. Going back to our example of the learner-driver with driving instructor, we can readily see that the learner is contributing to the motion of the car by bringing skills and resources to the situation which s/he already has. The instructor provides additional resources and skills (mostly in the form of speech: feeding back to the learner, providing sequences of steps, etc.) by which the entire activity, driving the car in a manner appropriate for public roads, is able to take place.

Clearly, the learner has the potential to drive the car in an appropriate manner. This potential is not realized until we introduce car, road and instructor into the picture. Note that the instructor is not simply giving the learner 'information' about how to drive. The timing of what the instructor says will be dependent on what the learner does. Their communication is essentially co-ordinative rather than informative.

Learning how to do something like driving a car can be said to depend to a greater or lesser extent on an ability to know when we are doing something wrong, and knowing when things 'don't feel right' and so on. In technical jargon these are sometimes referred to as 'auto-critical' skills. They occur all the time in daily life in our 'private language' or 'inner dialogue'. If you ever go to put salt in your tea, divide when you mean to multiply, or throw away the banana and go to eat the skin you will be thankful for the mobilization of auto-critical skills.

Part of what happens when we acquire a skill for ourselves, rather than always needing an instructor to sit next to us for example, is that we learn to take over the self-critical function for which previously someone else took responsibility. As babies get more competent and less blundering in their attempts to feed themselves we tend not to congratulate them just for picking up a spoon. We raise the stakes as time goes on and the baby takes more responsibility for the whole sequence of events from putting the spoon into the bowl and carrying it and the food triumphantly to the mouth.

Now all this may read as if bits of knowledge are going into people's, or babies', heads. Above we have written of children with learning difficulties and indicated that the methods we use to look at the nature of the difficulties often lead us to think that the interior of heads is bound to be the place where things are going wrong. Self-critical functions might also be thought to be about what goes on inside heads. We cannot take this argument further here, but we can indicate that caution is required. Experimental methods, like methods for looking at children with difficulties, can also lead us to focus too intently on individual children and not on the nature of the contexts in which they are operating. Below we quote studies that we feel are biased toward looking at children in such ways, but which nevertheless say interesting things.

Brown and Ferrara[27] report that 'retarded performers' tend to lack a variety of 'systematic data-gathering, checking, monitoring and self-regulatory mechanisms'. They also discuss an Israeli project (Feuerstein, 1980) in which retarded performers engaged in cooperative problem-solving. Feuerstein called the process he set up 'Instrumental Enrichment' (IE). He wanted supportive teachers to assist children in solving problems, but the assistance was directed at making a child aware of her/his own learning process. In effect, this meant that Feuerstein saw these children as having potential to solve problems alone if they could take on responsibility for regulating and criticizing themselves while working on an activity. Initially a teacher would take on this function and gradually the child would take it over.

Taking it over means making a skill subject to one's own volition. Without wilfully taking charge of activities, Feuerstein believes, children do not become motivated to grasp the nature of problems as wholes. This makes it more difficult to gain insights into problems. Grasping problems as wholes and insightfulness seem to be central to the ability to transfer skills learnt in one context to another context. Transfer appears to depend more on knowing about contexts and knowing how to judge how one feels when solving problems in those contexts than on having precise skills for terminating a problem or coming up with a final solution.

This is important to remember in mathematics education. Finishing off a problem or activity with the flourish of a solution takes little time. Understanding the problem and making sense of it relies on less automatic and more time-consuming work. Furthermore, remember that many mathematical activities are open-ended or admit of several solutions.

Within cooperative learning we can identify forms of interaction which assist the learner in becoming confident. Another rich field of work shows us cooperative learning from different angles. If you consider again the game we observed above you will recall that the nature of the game was itself obscure for all the children. None of them had played it before, yet they generated ideas about what the object of the game was and how they would recognize winning when it happened to them. We can apply the notion of auto-critical skills to this. The children were monitoring themselves and each other quite skilfully. But notice there was no-one who had a 'correct' perspective of the game from which to be critical.

In the context of experimental research it is clear that peer interaction around a common problem will facilitate learning in individuals.[28] Why this happens is a matter of further enquiry. There have been suggestions[29] that when, for example, two children work on a problem together each provides different perspectives on the situation which then come into conflict. New insights occur in the resolution of this conflict. The nature and value of this conflict is an academic argument. It is difficult to separate the children and the problem from the experiment as a contrived setting (*see*, for example

Shotter, 1984).[30] Furthermore, the design of these experiments tends to make the individual a focus of attention, strangely enough, at the expense of looking more intently at the 'social' (*see* Vass and Merttens, 1987).[31]

In these experimental studies it seems that conflict may enable children to *objectify* their perspectives which they then use as hooks on which the problem may hang while they examine it further. Conflict allows the mutual construction of a shared understanding of the problem in hand. In experimental situations where the problems under review are the traditional Piagetian ones of conservation and so on, it has to be said that the performance of pairs of children may be inferior to individuals. However, we have already implied that no two children bring precisely the same resources to a situation, and very often one of two children will take a leading role in which the second child may 'hitch a lift' rather than help directly.

For as Perret-Clermont *et al.*[32] say '... it has been repeatedly observed that situations where individuals must coordinate their actions with one another lead them to produce new cognitive co-ordination of higher competence. Subsequent individual performances demonstrate that these collectively attained competences are consequently interiorized by the actors'. In other words, in sharing, the children learned.

Where more-able peers are working cooperatively with less-able peers, Wood writes that the more knowledgeable 'can assist them in organizing their activities, by reducing uncertainty; breaking down a complex task into more manageable steps or stages'.[38] Professor Wood discusses the notion of scaffolding in the learning process. The image of a social 'scaffold' for our discussion is a compelling one. In order to construct an effective scaffold within which to undertake a task there are several important considerations to bear in mind. Firstly, interaction between children should be goal-directed. Even if temporary goals are constructed it is better than no goals: the scaffolding of activities is built around and toward the goal in question. Wood comments that children can recognize goals before they know precisely how to achieve them. He goes as far as saying that 'a learner's incomplete understanding of what he or she is shown and told ... is a vital basis for learning through instruction'.

It seems that the notion of 'transfer' is one which presents us with problems. The idea of transfer implies that knowledge goes into children's heads and that our job is to enable them to extract it in different situations. Yet many of the studies quoted above indicate that the way unfixed, even incomplete, knowledge engages with unfixed situations should be looked at more closely.

We have now come a long way in considering the details of cooperation in learning. But we have not probed very much into the nature of instruction. We have mentioned how speech can coordinate activities, and we have mentioned that in group contexts there will be an inequality in who possesses what speech or intellectual resources. Before leaving the subject entirely and returning to a consideration of IMPACT processes we should examine briefly 'peer-tutoring' programmes.

Peer Tutoring

Where children or adults are specifically trained or instructed to tutor others according to given criteria we are in the realm of peer tutoring. The word peer, it seems, is extended to include tutors of a wide range of ages for any specified child. More accurately we should refer to situations in which children are tutoring adults, or vice versa, as *cross-age tutoring*. There are, however, great similarities in approach and effect in the two types. Adult literacy and numeracy schemes could be cited as examples of peer tutoring. But these are special cases when we consider what those working in this field think that peer tutoring does.

Keith Topping[34] provides a concise history of peer tutoring in his *Peer Tutoring Handbook*. The development of interest in the topic emerges from practical considerations when teachers were confronted with very large numbers of children. Nevertheless, those that came to employ peer tutoring as an educational technique ceaselessly evangelized it for pedagogical reasons. The claim, and a substantially verified one, is that peer tutoring brings benefit to the tutee but often more benefit to the tutor.

1989 saw the publication of Goodlad and Hirst's book with a title that sums up the philosophy of the whole enterprise: *Peer Tutoring: A Guide to Learning by Teaching*.[35] In many ways the activity of tutoring is a branch of cooperative learning. The difference is that the tutor is singled out as someone who *has* the knowledge and who will be, at least initially, controlling the learning situation. This is a social signal that is not lost on the tutor. The tutor is being valued by being singled out and no doubt this is motivating. Goodlad and Hirst classify the benefits to tutors under the following headings:

Benefits to Tutors
1. Tutors develop their sense of personal adequacy.
2. Tutors find a meaningful use of the subject-matter of their studies.
3. Tutors reinforce their knowledge of fundamentals.
4. Tutors, in the adult role and with the status of the teacher, experience being part of a productive society.
5. Tutors develop insight into the teaching/learning process and can cooperate better with their own teachers.

Benefits to Tutees
1. Tutees receive individualized instruction.
2. Tutees receive more teaching.
3. Tutees may respond better to their peers than to their teachers.
4. Tutees can receive companionship from tutors.

They also distinguish unstructured from structured peer tutoring. In the former the tutor has personal responsibility for setting up the situation and goals of the tutoring session and is free to arrange it in his or her own way. This maximizes the opportunity for the tutor to accommodate his or her teaching to the responsiveness of the tutee. This form of tutoring tends to be more successful when there is a large difference in age between tutor and tutee.

Structured peer tutoring relies on closely controlled procedures. Tutors with little background in the task can administer these procedures with benefits to the tutee. There is less room for reacting to contingencies that occur during activities which Wood has pointed out seems to be important for cooperative learning in general. But there is room for a semi-structured form of tutoring in which the advantages of both sorts can be harnessed.

Home and School as Situations for Learning

In the Introduction we picked out the notion that a central feature of IMPACT was the role typically adopted by the child in the home. At home the child is seen as a source of 'tutelage'.

Homes and schools give us very different situations or contexts for doing activities and learning. As we said much earlier, different contexts will call from us different linguistic resources and techniques for managing problems. Since we have identified that doing maths means organizing, setting up, coordinating, choosing strategies, etc., not only for ourselves but possibly for others too, we should briefly say what effect homes as opposed to schools might have on such things.

As part of our pilot research we made several video-tape recordings in homes and in schools. Observing the same children in both places we were struck by the differences in approach and manner of execution of tasks — but most specifically in the ways very young children enlisted help from adults and used their tutelage in the pursuit of their intentions.

One nursery child, Carol aged 4, had been engaged in the classroom looking for things which were bigger and smaller than her pencil. She needed to collect four objects bigger and four smaller. The classroom was bustling with conversation and activity and the teacher was working with four other children. Carol went off to retrieve various objects and four times brought back objects of the same size as her pencil though she was able to compare them on route. On the first two occasions the teacher had said simply 'no, those are not bigger or smaller'. On the next two the teacher attempted to make sure Carol knew what was bigger and smaller. Carol behaved as if she needed to be taught from scratch. While looking around the room she appeared lost at times and then excited at coming across objects which she would take back to the teacher for comment. The teacher felt that Carol was picking up objects at random and just trying them out.

At home, Carol presented differently. The IMPACT activity was to make a small box and find three things that would fit inside it. The activity required some of the skills which she had been using in class that week. While mum acted as a sounding board, Carol set about the task with some determination and energy. She set up and organized the entire task, giving detailed instructions to her mother about the box and where she was to find the things for putting in it. Mum was quite clearly the assistant who had to be corrected when 'mistakes' according to Carol had been made. Carol found two objects to go in the box and her mother pointed out three were required. This incident was the only one in which Carol did not take the initiative.

This change of orientation by Carol echoes findings in other studies which have looked at children at home and in school. The Bristol Language Development Project is a rich source of data on this subject and much of it is published by Gordon Wells. He tells us[36] that in one sample of 32 children who were recorded at home and in school they initiated conversation in 63.6 per cent of occasions at home but only 23 per cent at school. At home 12.7 per cent of utterances were questions, but only 4 per cent at school. Clearly the opportunities for certain forms of social engagement are different in home and school. The school is not 'coming off badly' in these figures. If two parents had 30 children the language opportunities for any individual child might be very much worse than in school!

During our pilot work on IMPACT we discovered that Siletti-speaking children (Siletti has no written form) were going home and translating the instructions they had acquired in school into their parents' mother tongue. The entire activity would then be conducted in their parents' language. The results of what they had done were translated back into English in the classroom. We found a greater sibling involvement with these children than direct parental involvement.

The IMPACT Process

In the light of what we have said so far we can take a less naïve look at the IMPACT process. We should reiterate that IMPACT did not start only as a result of the research we have been talking about. We hope that research gives deeper insight into the routine practice that we engage in every day. There will be further discussion of this in the evaluation chapter of this book (chapter 6).

Types of IMPACT Task

Since IMPACT home-tasks are constructed out of on-going classwork they will vary in terms of curricular and educational content. Sometimes a teacher may want a child simply to gather data for subsequent class use; on another occasion there will be problem-solving or investigational work. This means that each type of task will have different implications for the type of interactions that children, their peers and parents engage in.

With respect to the issues we have been addressing in the aforegoing discussion of cooperative learning, we can see that some activities will entail children going home and formulating questions for their parents so that a class survey on, for instance, tea and coffee drinking may be carried out. On other occasions parents and children will be engaged on experiments (such as finding and predicting what sort of objects found in the home will float and which will sink). The latter type of task will typically require more of the 'coordinative' types of interaction between parent and child. Other tasks will require parent and child to devise strategies for winning board games in less than six turns. And still other tasks will ask parents and children to negotiate their way through an extension of an activity that children have already carried out in a different form in class.

When we look at the tasks from the point of view of research they will fall into one set of categories. Nevertheless they also fall into a set of more everyday categories that teachers commonly use, although we should say there is no direct 'mapping' of the two sets of categories.

Tasks to be carried out at home, we argue, should not be valued or even seen in isolation. As we mentioned in the introduction, the home task forms a part, sometimes a small part and at other times a larger part, of what we refer to as an *activity cycle*.

The Activity Cycle

The cycle represents a part of on-going classwork. It always has 3 phases:

Phase 1 — Preparation in the Classroom
During the week before the home task is taken home, the class will be engaged on work that will contribute to the design or selection of the task. Preparation can consist of specific class activities, including work on the maths topic, or work integrated with other curricular areas. There is also a 'rug' session in which the class is addressed as a whole and given a briefing about the home task by the teacher. The significance of any particular form of preparation will depend on the nature of the task as a central focus for parent–child interaction. For example, when surveys are carried out at home it is more likely that the rug session will be a significant form of preparation since the teacher will use the opportunity to address the children according to a survey format.

Phase 2 — The Home Task
The teacher, having focused on some particular aspect of concurrent work, will either select or draw up a task based on that work. The teacher will have in mind:
1 how the home may be utilized in the extension of classwork,
2 the type of effects and outcomes that the task might produce, and
3 how these effects and outcomes might be woven into and developed in subsequent classwork in accordance with the concurrent teaching plan.

Phase 3 — The Follow-up
Further classwork:

Sharing Maths Cultures

1. is adjusted to the outcomes of the work at home,
2. uses any content or material from the home in the classwork, and
3. utilizes feedback from the home by taking it into account when designing further classroom activities.

Other aspects of follow-up include the mounting of displays and opportunities for integrating activities across the curriculum.

Tutoring at Home

By asking children to become tutors in the home we are requiring them to be aware of the shape of the task, to search for its possible goals and outcomes and to communicate this to parents in the coordination of the work at home. School language, ideas and metaphors mix with those typically used in the home. If the approach to maths differs markedly at home from that taken at school we might look to the interactions of children and parents to 'solve', as it were, categorical confusions by recourse to more active and coordinative work. It would be very unusual for the parent of a six-year-old to expect a 'lecture' from the latter about the nature and meaning of the task (although it has happened at least once in our experience!). When confusion or misunderstanding arises in the initial moments of doing an activity at home it is more usual to hear 'Hmm . . . let's have a look at this . . .' or '. . . let's try this another way. . .'.

We have many examples where initial misunderstanding of the task has meant that parent and child have done something slightly differently, or even re-designed the task. Surely this is a good thing. We do not want parents to feel their children are being given self-contained products in these home tasks. The value of a misunderstood task must not be underestimated! With humility we may have to accept that the cereal packet is sometimes more nutritious than the contents. Adults have had more experience than children in handling themselves when they feel they are in error. It is of inestimable value for children to pick up a sense of how we deal with ourselves in a situation where we appear to be making mistakes (remember how important auto-critical skills are).

Bringing the Home to School

By displaying work done at home in the classroom you make the latter a different sort of place for the children and the parents. Children can offer back in classwork and in class-talk, material over which they feel some unique expertise. There will be stories about this material which children may tell in class. Being able to see parts of the classwork in the light of quite personal events at home may help the children to objectify aspects of current tasks.

Widening the Institutional Base of Teaching

Much of what we have to say exists as opportunity for schools and homes to grasp. Not all activities will be of the same quality, follow-up may not always be consistent, and preparation may not always adequately prepare. But IMPACT is not a controlled experiment. It is a widening of the cultural and institutional base of education by using elements which are already there. These elements are the skills which we all, parents teachers and children, use every day. Education as an institution must explore ways of bringing all these opportunities together. Using energy to keep apart the two main worlds which children inhabit, school and home, is surely to our disadvantage. While our interests are invested in distancing school and home, we remove opportunity from ourselves to reflect on our practices perhaps more deeply, and certainly more insightfully.

Wood says (*ibid*) that our views about the children we teach are partly a product of the ways we are constrained to interact with them given the type of institutions within which we live and work. He goes on to say that,

> If we find ourselves dissatisfied with the interactions that take place in such institutions [that we have invented to bring teachers and learners together] measured against what we take to be the optimum contexts for learning, then we must question not simply the teacher's 'skills' but the *form* of the institution within which we expect these to be deployed. (Our emphasis).

If anything IMPACT is a means for taking a closer look, as a community, at the *form* of teaching and learning.

IMPACT and the National Curriculum

The major recent institutional change to the shape of education in England and Wales came in the form of the 1988 Education Reform Act (ERA) and in particular that bit of it known as the National Curriculum. ERA provides for major changes in the way schools are managed. But the National Curriculum will subtly alter the constraints within which we plan and conduct those interactions we have been talking about that foster teaching and learning.

We used to say that a version of IMPACT can be made to fit any school. Now it has to fit with the ERA as we pointed out in the introduction. ERA does increase the accountability of schools to parents. But this may take the form of schools sending *more* information home to parents. While greater opportunity now exists for parents to contribute their ideas to the education of their children, ERA tends to focus parental power toward school management. The basis of their information about the school is the aggregated assessment data that schools will be publishing as an index of their performance. Parents will also have detailed curriculum information about their own children set in the context of how other children are doing 'as a group'.

It has been said that the dominance of the 'expert-client' view of the school–home relationship will change into a 'producer–consumer' one. Our experience of parents in their association with schools is that they are far more interested in their own child's progress and in exploring how they may contribute to it than they are in acting as some sort of quality controllers on schools.

The National Curriculum does not necessarily force us to act in ways that will damage teacher–child interactions, although it will change some aspects of the role teachers now inhabit as floor-managers of the curriculum. The National Curriculum has altered the shape of the curriculum by prioritizing certain elements over others and by forcing us to consider assessment as an element inextricable from teaching and learning.

We feel that it is essential that teachers and parents collaborate within the framework of the curriculum. It may transpire that the ERA is a neutral instrument in the changes that will occur in the home–school relationship. While we can see that schools and parents can become further entrenched in their separateness by the provisions of ERA, we do not readily see that ERA should be held responsible for this. The relationships that parents and teachers can foster in the pursuit of greater educational opportunity for children, are attainable in practice within the legal framework of ERA.

The teacher has greater administrative responsibilities with respect to the new curriculum. We feel that their expertise might be usefully directed to setting up situations within which certain kinds of interactions can occur. With reflection and insight teachers might turn their management of the curriculum into the constructive guidance of others in the pursuit of curriculum objectives.

Most of all we would like IMPACT to be a starting point for schools and parents from which to discuss matters of education, learning, progression and the curriculum. We do not give assurances that such discussions will always be comfortable. We do not think that passing Acts of Parliament mean that the debates on education are in some way concluded.

National Curriculum and Mathematics

There were no great surprises when we saw the final shape of the mathematics curriculum. *What* we teach has changed very little. This is comforting and reassuring for teachers. But we may feel a little more anxious when we come to consider the contexts and situations within which maths must now be taught.

In effect the curriculum lists statements of attainment (which we discuss in greater detail in subsequent chapters). These statements of attainment are skills-based because they need to be easily assessable. There is nothing wrong with this in itself even though many rightly warn that having skills as *educational objectives* is a recipe for teaching by rote-learning.

We have seen in this chapter that skills of all kinds are very important in the acquisition of competences we value as a society, and in human development in general. But we have also seen that the circumstances in which children (and the rest of us) develop and use skills are far more complicated than one might think. Furthermore, it could be strongly argued that organizing social contexts — children sitting around a table for example — primarily for ease of assessment purposes, is counter-productive to teaching and learning.

If the mathematics curriculum is to provide opportunities for children to acquire competences for everyday life we have to pay attention to the skills that belong to making sense of the problems we are given and in being able to formulate questions of our own. Knowing how to share out cake at a party; how much water needs to go in a kettle to make two cups of tea; knowing how to estimate running totals of prices in a supermarket trolley; knowing how to plan a route and how much to pack — all these skills require people to have been allowed an opportunity to gain insight into the situations in which we find such things important. These skills have less to do with precision calculation (which has its own value in its own context) and more to do with planning and getting on with our lives — we are sure Plato would approve.

Chapter 3

Getting Started

Cape Diem!
with thanks to the Dead Poets' Society
(and Robin Aldridge)

National Curriculum Process

We often say that IMPACT is more like a pint of beer than a plate of chips! It has to take the shape of the context into which it fits. This means that what IMPACT looks like in one school will not be the same as in another school. However, since the Education Reform Act, IMPACT like every other intiative has had to fit the context of the National Curriculum. When explaining the IMPACT process, it helps if we can share the same picture of the National Curriculum framework.

The National Curriculum has not so much changed what we do by writing out the content of the curriculum in Attainment Targets. It has radically altered the basis of the curriculum by arranging its content into ten levels of attainment. It is the fact that the curriculum is now organized primarily for assessment purposes that makes the difference. It means that there is now no piece of curriculum content which is not also a criterion for assessment.

The process illustrated in Figure 3.1 is now fairly well established within the routine of the school organization. Most teachers are planning their curriculum, usually starting with a topic. They then consult the Attainment Targets and Programmes of Study and decide which ones are going to be the focus of attention this term or half term. Keeping records is normally considered along-side the planning, so that the headings under which children's progress can be recorded are given by the plan.

Changes brought about by the National Curriculum

Part of the effect of the National Curriculum so far has been to ensure that far more time and energy is given over than previously to the process of planning and to particular procedures for keeping detailed records on each child. Furthermore, many teachers are approaching at least part of the curricu-

Getting Started

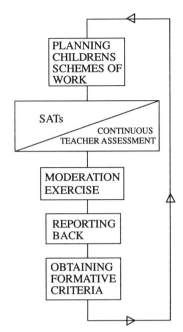

NATIONAL CURRICULUM PROCESS

lum in an integrated fashion. It is impossible to cover all the required attainment targets consecutively. This is seen by many as good, and it is expected that these trends will result in some improvements to primary practice in general.

However, there are also those who are sceptical about the effect of the National Curriculum. Many teachers have voiced their concern that teaching is now seen as a pre-planned set of procedures, where very little or nothing is left to the teacher's responding intuitively to the needs or desires of the children. The requirement for written and detailed plans in advance appears to mean less *ad hoc* response and more rigid protocols, fewer moments of change or adaptability, and more sticking to timetables. For some, the National Curriculum signals a move away from child-centred education. We need to be aware of such changes as we perceive them to take place.

IMPACT relies upon planning in advance. But it builds in, as part of the process, an opportunity for the teachers to respond to what has happened at home. IMPACT therefore imports a degree of flexibility into what can seem an otherwise over-rigid timetable.

IMPACT and School Policy

Each school will have considered its own National Curriculum Development Plan and will have made decisions about how the curriculum is to be delivered

Sharing Maths Cultures

throughout the school. This will include such things as how continuity is to be maintained, whether or not any commercially produced schemes are to be used, what provisions there are to ensure a lack of repetition in terms of topic work from class to class and year to year and so on.

IMPACT Process

IMPACT is a process and not a commercial scheme. It does rely upon the selection or design of materials but these are neither the source nor the end point of the process. IMPACT, basically, simply involves the sending home of weekly/fortnightly activities and allowing what happens at home to construct at least a part of the subsequent week's classwork. The whole process can be depicted in a diagram.

IMPACT fits with any form of classroom organization, teaching style or curriculum structure. It is not dependent upon particular styles of practice. It is true to say that some styles of teaching are more amenable to an 'IMPACT-type' approach than others. It is also the case that IMPACT will mean some alteration to certain ways of organizing the classroom. But in principle, there is no reason why any teacher in any class in any school in any catchment area should not use IMPACT.

IMPACT Effects

This is not to imply that IMPACT has no effect upon a school's curriculum policy. The idea that each week's or fortnight's work is at least in part constructed as a follow-up to an activity done at home has many implications for the maths curriculum. It asks that teachers make explicit and continuous connections between the maths they do in the classroom and the maths that

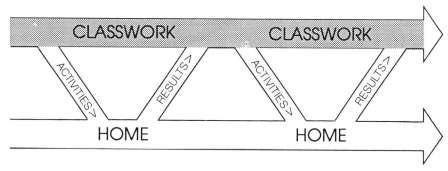

IMPACT PROCESS

Getting Started

children meet elsewhere. It involves an acceptance that what is done at home by parents and children together will, in practical and immediate ways, affect the classwork. In addition, it means that 'home' maths, unofficial or informal methods and non-scholarly approaches, will find a place alongside more conventional procedures.

IMPACT may also involve a revision of the school's policy regarding parental involvement. The idea that parents should be seen as partners in their children's learning is not new, but the notion of a direct parental input into the classroom curriculum is not normally something which teachers or governors have considered. Since IMPACT will involve an alteration in the patterns of contact with parents, by meetings, informal discussions and written comments, it is important to make reference to the proposed changes in the school policy document.

Welcoming Parents

A school may have a general policy of involving parents, but this may have traditionally amounted to the odd parents' evening and a small amount of help in the classroom mixing the paint and listening to the slow readers. IMPACT will mean that the parents will be sharing regular activities with their child. They will be commenting upon these through diaries or comment sheets, and will then be talking to their child's teacher, either informally before or after class, or in meetings about IMPACT.

This represents not so much a change of school policy as a revision of what it means in practice. It is this that needs to be spelt out in the policy document, through a brief outline of what is meant by 'involving parents', for example:

> This school involves parents in a variety of ways:
> through shared reading and maths activities (IMPACT) we expect parents to participate directly in helping their child's learning. The school welcomes parents' comments and suggestions on these activities;
> through inviting any parent to come and help in the classrooms at particular times during the week. Parents can help in a variety of ways, reading with the children, cooking, playing maths games, or learning alongside the children as they attempt to master a particular skill, e.g. on the computer;
> through informal and formal meetings, at which matters of interest to parents and teachers are discussed.

and so on . . .

It is by spelling out how parents are invited to participate in their children's education that the school policy demonstrates not only an intention to involve parents, but the practical means by which this is to happen.

Sharing Maths Cultures

Staffroom Decisions: Things to Consider

Several staff discussions are necessary before a school starts up IMPACT. The whole staff needs to consider the implications of involving parents in this way. It is not necessary that every teacher is equally enthusiastic, but it is necessary that they are all equally well informed. Sending home weekly activities and following them up in the subsequent week's classwork can sound like a lot of work at first, and there will be some teachers who have tried a number of new initiatives in the past and who prefer to sit on the fence until anything new has proved its worth.

Things which need to be considered by the whole staff include:
- which teachers?
- which classes?
- how frequently, weekly or fortnightly, IMPACT activities should go home?
- how IMPACT is to be resourced, in terms of photo-copiers or banda-machines, paper, etc.
- whether it is to be coordinated throughout the whole school, and if so, by whom?

We will consider each of these points.

Which Teachers?

Enthusiasm — an important factor? During five years of living with IMPACT, we found that one of the important factors in determining how many children and parents respond to an initiative like IMPACT and actually share the activity each week, appears to be the enthusiasm with which the teacher sends the activity home. Bearing this in mind, it might seem obvious that it is best for a variety of other reasons as well to start IMPACT in a school with only those teachers who are most keen on the idea. However, those teachers on IMPACT who proved to be most enthusiastic were not necessarily those who were that way at the start of the project or when they were first told about it.

One teacher, only a couple of years off retirement, began her conversation with us by saying, 'You may as well know that I didn't want to be involved in this ... I'm not very keen at all. ...' After such a beginning, we were not optimistic about the chances of IMPACT going well in her classroom. But we would have been quite wrong if we had made any firm prediction because this lady, a caring and responsive teacher, found to her (and our) surprise that she and the children got a tremendous amount out of the shared maths activities. In fact, she wrote after two years of IMPACT that she was really pleased to have been involved and that she could not have imagined how well it would work in the classroom. And it is a tribute to her open-mindedness that she and the children gained as they did.

IMPACT — overcoming inertia! IMPACT involves not so much working harder, as working in a different way. Change is always hard, and perhaps is harder for those of us who are more set in our ways than for those new to teaching. But although actually changing how we work requires effort, once we have set up the new routines, IMPACT can mean that some aspects of teaching are actually easier. We have been often quoted as saying that getting IMPACT started is like pushing a ten-ton lorry. It takes an immense amount of effort to get it moving in the first place, but once it is under way, we can all hitch a ride on the back.

No mathematical genius needed! It is not important that a teacher who is keen to send home weekly maths activities is an 'excellent' maths teacher, or that she considers maths to be one of her strengths. Many teachers' own experiences of being taught maths, both at school and at college, left them without much confidence or enthusiasm for the subject. IMPACT can be a useful means of generating enthusiasm for aspects of maths which are sometimes difficult to organize in a busy classroom. For example, if a teacher finds that she relies heavily upon a workbook-based commercial scheme, and finds it hard to see where and how the children should use and apply their mathematical skills, then IMPACT can provide a means of structuring this type of activity.

Don't wait until the maths curriculum is 'right'! A teacher once told us he would do IMPACT but he felt that he ought to wait until he got his classroom maths curriculum sorted out before involving the parents. A colleague responded to this by remarking that it seemed perfectly reasonable not to involve the parents in the maths curriculum if he didn't feel ready, provided that he kept the children out of it as well!

We will never reach the point at which we have finally got the teaching of maths 'right'. We have to accept that teaching is part of a constant process of trial and improvement. This means that we are always going to be asking the parents to participate in something less than perfect, unfinished, even messy. But this is not only how education is, it is how learning is. And we are asking the parents — with their children — simply to share in some part of the classroom activities and to tell us about it so that we can judge how it went and how well their child is doing.

Which Classes?

There are several criteria which might be applied when considering which classes are most suitable for starting IMPACT.

Age of children. We have found that children of all ages can gain a great deal from taking home maths activities on a regular basis and sharing them with their parents. However a number of points need to be borne in mind.

1. Running IMPACT in the nursery or reception class, or even in Year 1 (middle infants), means that for this age children the activities are usually sent fortnightly rather than weekly. It is very difficult to complete all the

necessary follow-up activities in a week with groups of children of this age.
2. Parental interest in their children's learning of maths tends to lag a little behind interest in reading. When the children first come to school, the parents are concerned that their child settles and is happy, that they don't cry at lunch time or be lonely in the playground. Once the child has been at school a while, then concern about learning to read starts to be felt. It is only when the child reaches top infants, Year 2, that parents, as a general rule, start to think, 'Why can't they work out the change?' or 'Shouldn't they be telling the time by now?' So, we have noticed that Years 2, 3 and 4 are ideal times to start IMPACT since this is a time when it seems to be particularly the subject of parental concern.
3. We are looking to IMPACT to affect the attitudes of both parents and children to maths. Since we also want sharing IMPACT activities to be a positive experience, it helps if the children have not formed negative attitudes towards either maths or the idea of homework. Top juniors, Years 5 and 6, can sometimes have already decided that sharing anything from school with parents is a bad idea. Street credibility depends upon a notion of 'school = work = bad' and 'home = play = good' and never the twain shall meet. We are hoping that as IMPACT infants progress up the school into the upper juniors, they will continue to regard the taking home and sharing of weekly activities as simply a normal part of school life.

Parallel classes. An organizational matter that the head teacher has to bear in mind when deciding which classes should start IMPACT is that it is difficult for parents to understand why one class of top infants is doing IMPACT if a parallel class of the same age children is not doing so. The parents of children in the class not 'IMPACTing' find it hard to believe that their children are not losing out. This does not automatically mean that all the parallel classes have to start IMPACT at once. But it does mean that the school has to reassure parents that working in this way is now whole school policy, and that eventually, all the classes will be involved. Parents, like anyone else, can see that it is often advisable to stagger the start of an initiative so as to give each class the maximum possible support while the routines are new and unfamiliar.

'I've started so I'll finish . . .' Another criteria to bear in mind when selecting a starting class for IMPACT is that once a cohort of children and their families have got accustomed to sharing activities in an IMPACT way, then it is very difficult to stop them doing so. When the class moves up at the end of the year, the parents may create a great deal of fuss if IMPACT suddenly stops being part of routine practice. The argument that Teacher 'X' does IMPACT whereas Teacher 'Y' does not, is hardly going to endear Teacher 'Y' to the parents at the start of the year. If the parents have been encouraged to participate in their children's learning of maths they are hardly likely to agree that such practices have suddenly ceased to be important.

IMPACT: How often?

There is no one set of procedures or routines on IMPACT which we know of that will suit everyone. In fact, there are almost as many changes in the detail of day-to-day running as there are schools — or even classes — on IMPACT. All that is possible is to share some of the experiences teachers have reported over the last few years.

IMPACT is mostly sent home once a week as a matter of routine. However, there are exceptions to this:

1 Teachers in the nursery or reception class tend to find that once a week simply does not allow the time for adequate follow-up work to be done. We shall be emphasizing the crucial importance of the subsequent week's classwork following any IMPACT activity. With very young children it is not possible to allow them the time to share what they have done and work in groups if there is only a week between each activity. A whole class of children will not all bring back their IMPACT on the first day of the week, and by the time all or most of the children have done the activity at home there may only be two days in which to follow it up. The experience of 100 per cent of the nursery teachers, and 90 per cent of the reception teachers, on IMPACT, is that a fortnight gives a much better time in which to operate the process.

2 Some teachers find that once a week is too often in the top juniors. If the children are asked to do a small weekly task, they are prone to regard it as homework and either to do it on their own, or to see it as a chore to be endured rather than as something enjoyable to be shared. If, however, the activity is more substantial and maybe takes them quite a few days to complete — a survey, a puzzle or an investigation of some sort — then this can be treated much more positively by children and parents alike. This is a case of little and often not being necessarily best!

3 Some head teachers, especially those in large schools in urban areas, can find that the resource implications of a whole school once-a-week policy are prohibitive. Since the nursery, reception and middle infant classes, as well as the third and fourth year juniors, arguably go better on a once a fortnightly basis, they decide that it is easiest — and most economic — to make the whole school the same. This has the advantage that parents know that their children, irrespective of class, bring home their shared maths fortnightly, and they can look out for it. It also allows teachers to stagger their IMPACT so that a parent with three children in the same school does not have to do three lots of IMPACT in any one week.

The most common practice for IMPACT is neither a rigid weekly nor fortnightly pattern but a mixture of the two. A teacher will send a data-collection activity, and this will generate so much classwork when the children's data all comes flooding back that another IMPACT activity will not be sent for a fortnight. However, if the next activity is something which requires the children to make something at home, or to practise a skill, then there is no reason why all the subsequent follow-up work cannot be completed within the week and another activity sent home.

Not less than once a fortnight . . . We have found that where shared maths activities are sent home less frequently than once a fortnight, the IMPACT routine destabilizes. The sense of the 'home maths' as an important part of the maths curriculum becomes obscure, and IMPACT simply becomes somewhat haphazard homework. In order to obtain the process outlined in Figure 3.2 it would seem that a minimum frequency of fortnightly activities must be maintained.

Which Day is IMPACT Day?

Although there cannot be said to be a standard day of the week on which IMPACT is always sent home, more schools send IMPACT at the weekend than at any other time. Both Thursday and Friday are popular days for giving out the weekend's IMPACT activity sheet. Some schools choose Thursday on the basis that it gives parents a chance to pop in on Friday to ask about any points which aren't clear to them, before they share the activity with their child over the weekend. Other schools opt for a Friday and expect to have a more staggered return at the beginning of the following week.

Coming back in dribs and drabs . . . No matter which day IMPACT is sent home, not all the children who respond in any one week will bring back their work on the same day. Typically, some children will turn up all bright and breezy on the Monday morning with their IMPACT, others will drift in later in the week with it. Certainly, parents' experiences on IMPACT demonstrate that peer pressure is a powerful force in getting children keen to do their IMPACT. We find that children who were far too busy playing or watching telly over the weekend to do their IMPACT come back home on a Monday evening and complain that 'Everyone else has done their IMPACT. . .). Whereupon parents rustle round to help!

The day of the week on which IMPACT is sent home does not seem to have any effect upon the number of children and parents taking part. We have schools on IMPACT who send the activity on a Tuesday, on a Monday or on a Wednesday, as well as the more common Friday or Thursday. It must suit the teacher, the children and the school. And sometimes the parents will put in a specific request in one of the follow-up meetings. Provided that the children are allowed a little leeway in terms of which day they bring the work back in, the number of IMPACT responses seems to be unaffected by this.

What about Resources?

Paper, and more paper . . . IMPACT relies upon the children taking home an activity to share with parents or siblings. Since the idea is that the child

Getting Started

should be able to explain the activity to whoever is helping, the sheet of paper on which the activity is outlined acts as a back-up. It provides a memory jogger; it sits on the kitchen table and reminds child and parent that IMPACT exists and is there to be done. However, children often complain that their parents don't listen to them; that they take the sheet and read it and then tell the child what to do. So we sometimes say that, in theory at least, the best IMPACT sheet is a blank sheet, because then the child has to do all the explaining and the parent has to listen!

In practice, each week or fortnight a piece of paper goes home describing what has to be done. Sometimes the activity is copied on to the back of the response sheet; sometimes there is a separate response diary. But in any event, at least one sheet of paper per child per week is required, which means that the major resource implications on IMPACT concern paper.

Person time. It is helpful if someone in the school eventually takes on the job of coordinating IMPACT. To start with, when perhaps only two or three classes are sending activities home, the teachers involved provide a support structure for each other and no one person occupies a coordinating role. However, as IMPACT expands throughout the school — and it is highly contagious! — it becomes necessary to have a person whose job it is to oversee IMPACT. It is important that there is continuity and that no repetition of activities occurs between one class and another. It is hard to explain to parents why a child has the same activity in the top infants that they had in the reception class and so it is crucial that records are kept and that someone is responsible for IMPACT throughout the whole school.

Sometimes the IMPACT coordinator in a school is the same as the maths coordinator. In other schools, this is not so and the person responsible for IMPACT may also be responsible for shared reading. We have not found that it makes any difference to effective coordination that the IMPACT person is a maths expert! The existence of IMPACT materials means that the job of deciding what activity to send can be one of selection rather than design or creation. (*See* Chapter 4.)

IMPACT materials. There is a series of IMPACT materials which schools can have which enable teachers to draw on the years of experience of other IMPACT teachers. These materials can be obtained from the IMPACT Project. A list of all the materials available and the address from which they can be obtained is at the back of the book. There are also sample sheets given in Chapters 4 and 5, and in the Appendix.

Financial management of IMPACT. As we mentioned earlier, the major resource needed to share maths activities at home is paper. Some schools have minimized the cost of this by combining the comment form for parents and children with the activity itself.

If a school asks its LEA to print several thousand comment forms, then these can prove very cheap indeed. The school then copies the activity on to the back of the response sheet, using a heat copier or a photocopier.

Parents' Associations. Since the paper is a neccesary resource to enable the teachers to involve the parents in a partnership in children's learning, it is

Sharing Maths Cultures

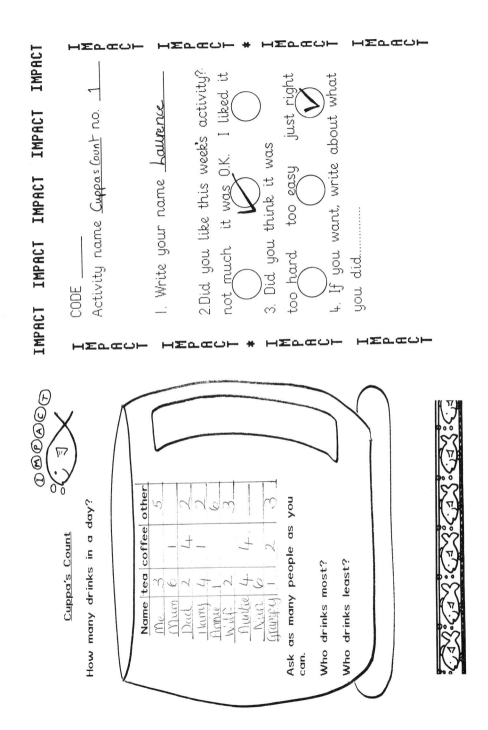

not inappropriate for the head teacher to ask the Parents' Association if they can assist by providing some of the paper needed. The extra cost does not seem much when looked at in isolation, but taken out of a school's budget it can be prohibitive. We are talking about something of the order of 15 reams of paper per year per 200 child school, which at current prices amounts to about £80. Most Parents' Associations are happy to spend any money they manage to raise on something which has such immediate relevance to the parents and children in the school.

Sponsorship. Another way of obtaining the necessary paper is to contact local firms and ask them for sponsorship. Some IMPACT schools have had considerable success in persuading local firms to supply the paper for, say, one term's IMPACT. Sometimes the firm will also fund the cost of printing the comment sheet onto the back of the paper as well. The best way of finding local firms is to use local knowledge. Sometimes a school governor can be of assistance here.

It is worth remembering that from time to time firms update their letting-headings and that when they do so, there is normally a wastage of the old headed-paper. It is worth alerting the local area CBI (Confederation of British Industry) and asking them to let firms know that schools are an excellent repository for old paper.

Sometimes it is possible to find sponsorship for materials other than paper. For example, many good maths games used on IMPACT involve dice, and it cannot necessarily be assumed that all homes will have a dice. Some teachers like to send dice home so that all the children get a chance to play such games. The cost of donating one dice to each child in the school is not high. This is the sort of thing which a local firm is sometimes quite keen to sponsor.

A question of priorities. IMPACT has to be said to be cheap at half the price, as they say. At the end of the day, it is a question of how we choose to allocate the resources available. For the price of one sheet of paper per child we can not only do extra maths — and more importantly, talk about the maths as we go — we can also involve the parents in something of real and personal interest to them.

Informing the Governors

Clearly it is of crucial importance to keep the governors informed of the steps that the school is taking to work in partnership with parents. IMPACT can be introduced to the governors as a new initiative in collaboration between teachers, children and parents. When discussing IMPACT with school governors, we have found that it helps to bear in mind the following points:

1 IMPACT is not a new maths scheme
It is important to emphasize that IMPACT is not some new-fangled maths

scheme that is going to replace whatever the teachers have been using up until now. We stress that the aim of IMPACT is really very simple — it is to involve the parents in their children's learning of maths. However the teachers organize their maths curriculum, they will continue in the same way, with the only difference being that they are asking the parents to share maths activities on a regular basis with their children.

The fact that IMPACT is not a commercial scheme means that each IMPACT activity that the teachers send home has been chosen, or designed, by the teachers to fit the maths they are doing at that time with their class. We need to stress that the sheets are not part of a scheme where a teacher just has to take the correct one for week 7 in term 2. The IMPACT activity brought home by Fred in Class 5 is unique to the work being done by Fred at that particular time. This makes IMPACT very different from any pre-designed schemes.

2 IMPACT parallels similar initiatives in shared reading

It is often best to introduce governors to the idea of IMPACT in the context of shared reading. With schemes like PACT and CAPER,[37] children choose a book and read it at home with a brother or sister or parent. Then the parent or sibling is asked to comment on how the reading went. Did the child enjoy it? Were some parts problematic? Did they have difficulties with certain words? With certain letters? Was the book too easy or boring?

IMPACT works in a similar fashion. The children bring home a maths activity. They share it with a parent or brother or sister. The helper is then asked to comment on how the activity went. Was it too easy or too hard? Do they think the child learned a lot or a little? Did they enjoy it? The main difference between IMPACT and shared reading is that frequently the children will bring something back into class as a result of the maths activity they do at home. But this does not alter the fact that IMPACT parallels in maths what the shared reading achieves in English.

3 Is IMPACT homework?

We often say that the main difference between IMPACT activities and conventional homework is that the former *are intended* to be shared. Traditional homework was something that you did on your own.... 'Go up to your room, dear, and get on with your homework quietly.' But the IMPACT activities are designed to be talked about, and done jointly with another person. They are not for the child to do on their own. Sometimes it would be impossible for a child to do the IMPACT activity on her/his own, and often it would not be easy. But it is never advisable, since we see the major benefit to children's learning as coming through the opportunity that IMPACT provides for maths discussion in the home.

4 IMPACT National Curriculum delivery

The governors, amongst others, are responsible for ensuring that the school is delivering the National Curriculum entitlement to each child. Therefore, they will be particularly concerned that the staff do not embark on anything which might adversely affect this delivery. Many governors are also only too aware that the implementation of the National Curriculum within the very short time allowed has been very stressful for many teachers. They are therefore extremely reluctant to see teachers taking on board anything by way of an 'extra'.

When talking to the governors it should be stressed that IMPACT provides one means of delivering the National Curriculum. Working 'in an IMPACT way' involves teachers in planning their curriculum, in keeping records and in listening and responding to parents' and children's comments. As we have often said, structured home involvement should not be seen as the icing on the cake, since in times of hardship people leave the cake un-iced. Rather, it should be seen as a different mix of cake. We are altering what goes into the cake, and this means changing how we deliver the curriculum, not ceasing to deliver the same maths.

5 IMPACT is not an 'experiment'

Five years ago, IMPACT might have been described as 'experimental' in that such a process had never been tried before in a substantial number of schools. However, after two pilot years in ILEA (1985–7), we knew that the process was one which worked.[38] That is to say, we knew that teachers could send weekly activities home which children could share with their parents and that the results of these could be used to help construct the following week's classwork. Three further years of IMPACT in other LEAS has confirmed this. Of course, no one can say exactly how IMPACT is going to go in any one school or any one class, any more than they can predict which children will gain most and in what the gains will precisely consist. But IMPACT is a tried and tested process, and should not be thought of or described as an experiment. We know it makes sense!

Governors are typically very supportive of IMPACT. Particularly in the context of the reporting-back requirements of the Education Reform Act, they are aware of the importance of building a constructive relationship between teachers and parents. The governors do need to be aware of the resource implications, and their advice may usefully be sought on the subject of funding the extra paper costs. Several governors may be parents of children in the school and it will be particularly interesting for them to be able to participate in a curriculum development initiative from the position of consumer as well as manager!

Summary

In this chapter we have described in some detail the processes involved in setting up IMPACT. Before any teacher starts to involve her/his children and parents in a system of structured home involvement in maths such as IMPACT, certain decisions need to be discussed and made. The whole staff need to consider the fact that, once children have got used to sharing a maths activity at home each week, they are unlikely to want to stop doing so. Also, there is the question of resources.

We have looked at how IMPACT fits with whole school policy. How do we decide which teacher, which class, how frequently to send activities, and upon which day of the week? We have considered the financial implications of IMPACT and finally thought about how best to keep the school governors informed.

It might seem that we anticipate problems where none exist. We felt it important to cover all contingencies and possible problems, since none have hitherto proved insurmountable. At the end of the day, the IMPACT process and its aim are both very simple. There is no good reason why any class teacher cannot start IMPACT in her class, fitting it in with her own personal teaching style and maths curriculum delivery.

Having set the stage, we are now ready to roll.

Chapter 4

IMPACT in the Classroom and the Home

The smile that you send out returns to you
                                            Indian Wisdom

The National Curriculum has meant some changes in how teachers set about planning their work and organizing their classrooms. Most of us are now starting to come to terms with what the new ways of working are going to involve and how we can best make them fit our own individual styles of teaching. Most of us will be approaching at least a part of the curriculum in an integrated way. It does not appear to be possible to cover the 227 statements of attainment at level 2 (in the Core subjects alone!) without teaching in a cross-curricula fashion. We shall involve the children in group work at least for part of their time, and we shall, of course be planning and recording as the ERA requires.

We feel that whilst working within the framework of the National Curriculum, the processes involved in IMPACT can assist teachers through allowing scope for responsiveness, and for a natural adaptation of plans and flexibility of approach. We can see how this works in practice if we work step by step through the National Curriculum process with IMPACT in mind.

Planning IMPACT

We plan IMPACT in conjunction with the rest of the maths curriculum. There is no sense in which the IMPACT activities can be treated separately since it is crucial for the success of the sharing of maths at home that the activities are an integral part of the classwork in maths.

Choosing a Topic

A good place to start in planning the maths is with the choice of a topic. Not all the maths will come out of this, but some of it will. Planning will enable us to see on which aspect of the maths curriculum we may wish to focus this term/half-term. We need to consider what topics the children have previously done, and also which parts of the Programmes of Study have yet to be addressed. For example, if the children have already done a topic on 'Ourselves', in which they did a fair amount of history and biology, as well as some maths and English, then it might make sense to choose a topic which will involve them in some physics, chemistry and geography.

Some topics can involve a fairly narrow focus upon one subject, whereas others have a broader coverage. At times it may be that a teacher will want to have more than one topic, perhaps one maths topic, and one in science or language, at the same time.

Example 1 Soap operas and the weather
One IMPACT teacher organized her term's work with fourth-year juniors through two main projects; one on 'TV and radio soap operas' and the other on 'the weather'. The first topics – soaps – was immensely popular with the children and involved them in a variety of work which included:
- Analysing how much air time (both TV and radio) is given over to 'soaps'. (Maths and English.)
- Discussing and analysing the structure of soap opera. What ingredients are necessary to make a successful soap? Why do some work so well? Can we produce an ideal 'soap' structure? (English.)
- Comparing the various current popular soap operas. This involved the drawing up of criteria upon which to base any comparison. It means exploring below the level of 'I like this one best', to what reasons can be given to justify a preference. (English.)
- Writing a review of any current soap opera. (English.)
- Calculating how long a script for a twenty-minute episode (radio) has to be. Starting the work towards structuring this. (Maths and English.)
- Working out the detailed recording arrangements including sound effects, more than one location, and tape recorder. (Science, CDT and maths.)

The subjects mainly covered in any one activity or set of activities are given in brackets.

The other topic, on weather, included:
- Making a chart over the period of the term showing the details of the weather for that period. This chart to include variations throughout the day. (Science and maths.)
- Researching, designing and making a device to measure rainfall. (Science, CDT and maths.)
- Performing an experiment to measure the atmospheric pressure in the second playground. Recording the results. (Science and maths.)
- Using a barometer to measure pressure changes.
- Discussing and researching what causes the weather. (Science and geography.)
- Making a model of part of the solar system to show the earth's relation to the sun, summer and winter. (Science, CDT and maths.)

These lists are not complete. There will, in any case, be a variety of pieces

IMPACT in the Classroom and the Home

of work which arises out of one of the aspects listed above and which has not been foreseen in the teacher's plan but which is none the less worth while.

The alternative approach is to have one broader topic out of which we hope to get some work across all the core areas and some of the foundation subjects.

Example 2 Noah's Ark
Another IMPACT teacher had chosen Noah's Ark as her topic with Top Infants (Year 2). Her list of the work she intended to get out of this topic included:
- Telling the story of Noah's ark, using the language of the Bible and other re-writes. Several versions to be used. (English.)
- Discussing the idea of having to save animals from a flood. Which animals would not need saving, and why? (English and science.)
- Looking at, listing and naming as many types of animal and bird as possible. (English and science.)
- Sorting the animals into sets – mammals, birds, reptiles, amphibians, insects, etc. Discussing the criteria in simple terms for each set. (Science and maths.)
- Children to visit the local zoo. They are to make sketches (with labels) of as many animals as they can. They draw these sketches carefully onto sticky labels. (English, art and science.)
- Making large sets on the wall, to which the names and drawings of any animal can be added. These sets use the sketches drawn on sticky labels done at the zoo. (Maths and English.)
- Work to start on making a large classroom Ark. Children to research this studying pictures and discussing suitable materials and structures. (Science, CDT and maths.)
- Each child to focus upon one mammal. First they should research that animal by finding as many pictures of it as possible. Then, they can start a detailed drawing of their chosen animal. (English and art.)
- Each child to write a description of their animal. (English.)
- The children to start making a book in which they are going to write a story of what happened to their pair of animals when they came out of the ark. (Art, CDT and English.)

'Bread-and-Butter Maths'

Whatever approach a teacher takes, and whether or not she/he has one broader topic or two or three more focused projects, there will be an opportunity for teaching some maths through this work. The children will have an opportunity to use and apply their mathematical skills.

We now need to make the connections between the everyday and on-going maths work which does not come out of the topic (the 'bread-and-butter' maths), and the topic maths itself. If we do not, as teachers, take care to make these connections explicit then it is going to be very difficult for the children to see any relationship between the two. This will considerably lessen the effectiveness of both the topic work and the 'bread-and-butter' maths.

Linking the Classwork to the Home Context

Since children have difficulty in transferring mathematical skills from one context to another, it is important that teachers make as many connections as possible between one situation and another in the context of what children are doing. We can envisage three types of maths in the IMPACT classroom: topic maths (or maths which arises out of a project), bread-and-butter maths (possibly covered at least in part through the use of some commercial scheme), and the home maths – that is the maths which the child shares at home with a parent or sibling. Explicit connections need to be made between these three.

There is a two-way relationship between the topic maths and the bread-and-butter maths. The topic work should enable children to see the point of the latter and apply some of the specific skills they are learning in a more restricted and formal context. The bread-and-butter maths should attempt to

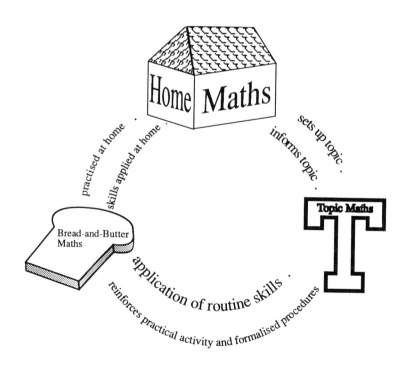

reinforce, and to some extent formalize, the practical maths that the children use in their topic work

The relationship between the home maths and the class maths is also two-way. The home maths can serve the function of enabling the child to practise and reinforce a skill acquired in the classroom. It can also provide another context, this time outside the classroom, in which the child can apply the skills she/he has learnt. On the other hand, the home maths can provide the data or stimulus needed to get a topic under way or to move it along from one area to another.

We may look to the home to serve three functions:
1. To provide some information or assistance with an aspect of the topic. This enables us to bring more of the 'outside' world into the classroom. It gives the topic a breadth which it cannot possess if all the information is generated from books or the school environment.
2. To enable the children to use and apply the mathematical skills which they learn in the classroom in a situation other than the class. This gives the children a valuable opportunity to transfer skills.
3. To allow children to explain a mathematical idea to someone else, to explore it and to practise skills in the context of a puzzle or a game.

It is important to have these functions in mind when selecting IMPACT activities and deciding what the children shall do at home. Some of the home maths should link with the topic work and other tasks should link with the bread-and-butter maths. There is no reason why this should be an even balance. At any one time, a teacher may lean more toward one than the other, but overall we need to ensure that both aspects of classroom maths are linked to the home context.

A Variety of IMPACT Activities

The range of activities which teachers have drafted and sent out over the last five years has reflected the three functions outlined above. We can categorize all IMPACT activities under three main headings:
1. Data collections.
2. Doing and making.
3. Games, puzzles and investigations.

This typology is depicted in Figure 4.2. It can be seen that the categories are not exclusive. It is perfectly possible for an activity to fall into two categories at once. Indeed, most data collections on IMPACT also involve doing or making, and many doing or making activities practise a skill.

From the point of view of the parents, it is helpful if the IMPACT activities demand a variety of skills, and if they involve different things each week. This is for two reasons. Firstly, it is boring if too many successive activities are asking children to do a different version of the same thing. Secondly, some of the IMPACT activities will be more readily recognizable as 'maths', from a parent's point of view. If a teacher sends six consecutive

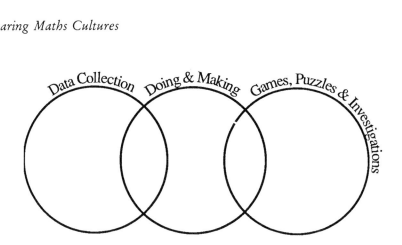

IMPACT Activity-types

activities, and not one of them looks anything like what a parent has come to regard as 'maths', then this will generate a great deal of anxiety. However, if a teacher sends home a number skills-practice game, followed by a data-handling activity, then the parents are inclined to look more sympathetically upon the idea that maths is broader than they imagined.

Study your topic plan. Having chosen a topic, the next task a teacher faces is to decide how this is to fit in with the bread-and-butter maths. Many teachers use a commercially produced scheme as the focus for their on-going, everyday maths. Other teachers have designed their own workcards. In any event, it is necessary to examine the maths which will come out of the topic in terms of which attainment targets and at which levels it is addressing.

Link with 'bread-and-butter' maths. Having made a list of the relevant attainment targets and levels, the next stage is to consider which other attainment targets you would be hoping to address in this term/half-term. Not all the maths will come out of the topic, except tangentially. We need to be certain that the children are getting some experience of all the aspects of number work which might not arise naturally from the topic. Even where a teacher is sticking fairly rigidly to a commercial scheme, it is possible to consider if:

- the children can work in groups through particular parts of the scheme. This will enable discussion and may maximize on the overlap between this work and the topic
- all the children can do certain parts of the scheme simply to reinforce some aspect of home or topic maths.

Fitting in IMPACT. Having studied the topic plan, and then having listed the parts of the scheme, or the aspects of her/his own system of 'bread-and-butter' maths that the children are going to cover, the teacher then has to consider which activities to send home.

Start with a data collection. Many teachers have found it extremely helpful in getting their topic off to a good start if they begin by sending home an activity which asks the children to find something out, or to provide some data of their own.

Example 1 Soap operas survey
The teacher doing the 'soaps' project, started the whole thing by asking the children to do an analysis of the percentage of time devoted to soaps over the weekend on any one channel of the TV or radio. The children did not *have* to watch all the soaps (!) but they did require access to a TV or radio guide in the evening or daily paper. This activity provoked, as it was intended to, a lot of discussion in the home about what 'counted' as a soap opera and why. It therefore led very neatly into the children's planned work on this topic.

Example 2 Noah's Ark – animal survey
The children doing the topic on Noah's Ark were asked at the beginning of the term to carry out a simple survey. They were given a large piece of paper on which were drawn two overlapping sets – one labelled 'fierce animals', the other labelled 'animals with four legs'. The children had to ask as many people as possible what their favourite animal was. They then had to record that animal in the correct place on the Venn diagram. They also had to note whose favourite it was, e.g. Tiger (Mum) in the intersection between the two sets. This activity got the project off to an excellent start since it focused the children's attention on a variety of types of animals and their different descriptions. The data were used back in the classroom to generate the first list of animals for the Ark, and to discuss their differences.

Get the children and parents doing or making. It is often difficult with thirty, or more, children in the class to give them the individual help they need. This is nowhere so apparent as when it comes to making something which requires a degree of technical skill which the children find demanding. IMPACT can offer an opportunity for children to make something at home where they are likely to have the assistance of a sibling or parent. This will enable a teacher's time to be devoted to those who did not get a chance to do the activity at home.

Example 1 The weather – make a helicopter
As a part of the work that the children were doing on the weather, they were looking at wind direction and wind strengths. The teacher sent home an activity which asked the children to make a 'helicopter' out of paper. Their helicopter required precise folding and had to be tried out for accuracy by dropping it from a height of about six feet and seeing if it would land on a marked-out square on the floor! The children had immense fun doing this activity at home. The teacher had provided them with card, as she would have done had they been making these helicopters in class. Many parents became extremely involved in the activity and several had tried to improve upon the original design. The discussions which several children had at home

Sharing Maths Cultures

Soap-Opera Reckoning!

Can you work out the amount of time devoted to Soap-operas on any ONE channel of the TV or radio over the next week?

You will have to make decisions about what is and isn't 'soap'!

Can you calculate the amount of time as a percentage of the whole week's viewing?

IMPACT in the Classroom and the Home

Animal Survey!

Ask as many people as possible what their favourite animal is. Write the name of the animal in the right place on the diagram below. Put the name of the person you asked in brackets beside it.

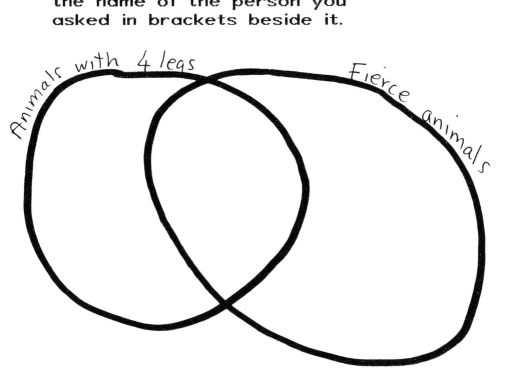

Sharing Maths Cultures

proved very informative when shared in the context of a whole class discussion of the problems.

Example 2   Noah's Ark – pet measuring!

The teacher doing the Noah's Ark project asked the children to measure a pet, or someone else's pet. They had to take a piece of string (supplied) and, with an adult's help, lay it from the tip of their pet's nose to the tip of its tail. Then they had to take another piece of string (also supplied) and wind it once around their pet's tummy, and cut off any excess. They were then asked to bring both pieces of string back into the classroom. This IMPACT activity was very useful in getting these young children to look closely at animals, and to consider their proportions. It formed the focus of a great deal of maths work back in the classroom in terms of direct and indirect comparison, and graphical representation.

How about skills practice? Some IMPACT activities do not initiate work in the classroom so much as reinforce something that has already been covered. This type of activity can vary from a fairly routine skills practice game where the child is throwing two dice and colouring in the totals on a board, to an investigation or puzzle, where they have to work out why something happens as it does.

Example 1   Soap operas – pocket money double-up!

The children doing a topic on soap operas spent some considerable time discussing the notion that riches appeared to be seen by many of the characters in the soaps as automatically conferring happiness upon their possessor. There was also a great deal of argument over what precisely 'rich' meant. Some children proved to be unclear about some of the large numbers involved, and the teacher decided to set them an IMPACT activity which would enable them to practise manipulating large numbers, and would also bring together some of their ideas about money and riches.

The IMPACT activity asked them to consult a parent as to which of the following alternatives they thought would be cheaper. They could either have £500 pocket money per week in the month of March, or they could have 1p on the first day, 2p on the second, 4p on the third, 8p on the fourth, 16p on the fifth and so on. They and the parent had to select their preferred option and they both then had to calculate how much pocket money the month would bring in both systems.

The activity involved considerable practise of doubling and adding. If the children – or their parents – were really sharp they noticed a pattern which saved them adding up all the days' pocket money in the end. But in any event they all got a surprise at the answers. The children had great fun with this activity, partly just from amazement at the size of the numbers involved. When they came back into class with their calculations, there was much creative dispute as to what the accurate answer was!

IMPACT in the Classroom and the Home

HELICOPTERS

Cut out the strip of paper.

Cut it down the centre to about half way, fold each tab in opposite directions (as in diagram)

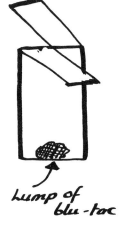

Weight the uncut end down some way
i.e. blue tac or a paper clip.

Test your helicopter to see if it will fly.

Try different paper, does it fly better?

Try changing the sides of the helicopter blades, What happens?

Make other changes to make your helicopter fly better.

What were they?

Bring your helicopters into school.

Sharing Maths Cultures

Measure a pet!

Can you find a pet to measure?

You will need to use the string provided.

You need to stretch the string from the tip of your pet's nose to the end of his or her tail. Then cut it off....the string NOT the tail!
Now stretch the string around the pet's tummy. Cut off the extra string.

Label both pieces of string – tummy and length.

Bring both pieces of string into class.

Draw your pet carefully.
Bring your drawing in.

IMPACT in the Classroom and the Home

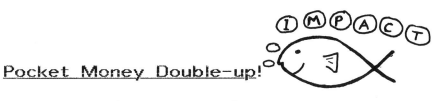

Pocket Money Double-up!

For your pocket money in March you have a choice:

You can either have £500 on each Saturday in the month... OR you can have 1p on the 1st day, 2p on the second day, 4p on the 3rd day, 8p on the 4th day, 16p on the 5th day.... and so on.

Which method will give you more money in March?

Find out how much you will get each way.

Bring ALL you sums back into class!

Example 2 Noah's Ark – even handed!

The children working on Noah's Ark were obviously doing a fair bit of work surrounding pairs and even numbers. In order to reinforce the idea of even numbers, after the children had spent quite a bit of time counting in twos in the class, the teachers sent home the following game called 'Even Handed!' Two people play, and they both put both hands behind their backs. Upon a count of three, they bring out their hands with a certain number of fingers (and thumbs!) standing upright and the rest folded down. They add up all the fingers standing up on all four hands. If the total is even, the 'Evens' person gets a point, if the total is odd, the 'Odds' person gets a point. Who is to be the 'Evens' person and who the 'Odds' person must be decided at the beginning of the game. The players play until the first person reaches ten points.

This game sounds very complicated, but in fact it is very easy, being based on the old game of 'paper, scissors, stone' which many of the children already knew. It raised a number of interesting questions as well as making very sure that the children knew their even and odd numbers. Some of the children felt that it was possible to fix the game by holding up a specific number of fingers. They were all puzzled as to what happened if no fingers were held up. Apart from the mathematical exploration in these questions, the children all did a large number of addition sums in a short time.

Planning IMPACT together

Overall, it is very helpful if the IMPACT activities are planned as a part of both the topic work and the bread-and-butter maths. The activities sent home are an integral part of the maths classwork. They can start a process, enrich an aspect of the work, or practise a skill. They can be seen as an opportunity to get something done which it would be difficult to get done in the classroom, either owing to the necessity of supervising thirty children or just because it *is* a classroom not a home. But IMPACT must be planned with the rest of the maths in order that it is a genuine part of that process by which the maths curriculum is delivered by the teacher to the children in the class.

The advantage of planning the maths to be shared at home as an integral part of the classroom curriculum is that at the same time as the IMPACT activity is planned, the activities which will follow it up in the class are also envisaged. This means that the home maths fits naturally and comfortably into the classwork rather than being an add-on.

The IMPACT Activity Cycle

One of the major differences between IMPACT and traditional homework is that in IMPACT the maths which is taken home comes out of and leads

IMPACT in the Classroom and the Home

Even-handed!

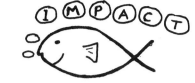

This is a game for two people, so you will need to find someone to play with.

Decide who is to be 'even' and who is to be 'odd'.

To play, you both put your hands behind your backs and, on a count of three, you both bring out both hands with some of the fingers on each hand standing up and some curled down.
Now add up the fingers on all four hands.

If the answer is even, the 'even' person gets the total as their score.

If the answer is odd, the 'odd' person gets the total as their score.

Play this ten times. Keep adding your scores. The one with the highest score at the end, wins.

Sharing Maths Cultures

back into the classwork. When the children bring their maths homework home from their secondary school there are four possibilities: the parent could do it (maybe!), the child could do it, they could do it together, or no one could do it! Whichever of these possibilities actually occurs, the class lesson the next day goes on regardless. The child may get into trouble for not doing his or her homework, but basically the structure of the lesson may remain unaffected.

With IMPACT this is not so. The classwork in the following week is, to a greater or lesser extent, dependent on the activity, the focus of some of the classwork. We think not of isolated 'home-activities' but rather of 'activity cycles'.

Preparation

The activity is always preceded by some preparatory work. This may be more implicit in the classwork than explicit, 'Here we are preparing for your IMPACT activity'. But it will involve at least a session with all the children – usually on the rug – where it is made very clear to them what exactly they have to do. IMPACT activities are never sent home 'cold'; that is to say, without the child having a really good idea about what the activity involves and what they have to do. Ideally the child should also be aware where the task fits with the work they are doing in class.

Cold = Bad! This is a very important point concerning IMPACT. If the child is not clear about what they have to do at home the nature of the task becomes something very different. If the teacher simply hands out an IMPACT sheet with an activity outlined on it, then the child cannot explain to the parent what has to be done. It will not be a case of the child 'instructing' or 'tutoring' the parent. Rather, it will be a case of the parent once again telling the child what they have to do. The quality of the experi-

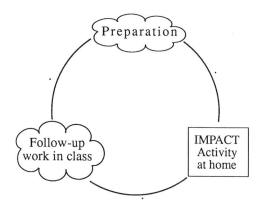

IMPACT Activity Cycle

ence, both from the point of view of the child's attitude to maths in general and to IMPACT in particular, and in terms of the enhancement of their learning, will be radically altered.

We often say that it is better not to send an activity in a particular week than to send one out 'cold', or without adequate preparation. The extent to which the children are firstly, clear about what they are going to do at home this week, and secondly enthusiastic about it, has a marked effect upon the number of responses. Remember also that the enthusiasm of the teacher tends to be a big factor in determining the number of responses from any one class.

Enthusiasm not Over-kill! The questions of what counts as adequate preparation for an IMPACT activity has been much discussed on the project. Some people suggested that we should produce official guidance to what 'adequate preparation' means. However, we have been reluctant to do this, believing that what adequate preparation looks like is determined by a number of factors including:
- age of children
- type of activity
- style of teaching approach
- classroom organization
- where it comes in the scheme of work.

Some activities require a great deal of preparation. The children will work together in groups, and they may do a number of tasks, worksheets or other activities which all in some way lead up to the activity being sent home. Other activities require little more than a session on the rug where the teacher outlines the task, perhaps demonstrates how it is to be done or shows something she/he made at home, and then gives the children the sheet to back it up.

Example 1 Pocket money double-up!

With the activity outlined above in which the children were asked to work out their pocket money for March by doubling-up, a great deal of preparation work went on. The children discussed what it would be like to be rich, how much a million pounds was in terms of what it would buy and how long you could live off it. Some groups of children calculated this and also worked out the interest it would make. Other children thought about the cost of living and worked out what sort of sums of money were involved in buying a house, running a car and so on. All this work was intended to familiarize children with large numbers and allow them to gain confidence in doing computations with large amounts. The teacher also gave children individual practice with a series of worksheets on doubling. There was a final session on the rug when the teacher outlined the problem and asked the children to offer some predictions as to which system would generate the most money and why.

Sharing Maths Cultures

Example 2 Pet measuring!
In the case of the pet measuring activity given above, the direct preparation consisted of an extended session with the whole class in which the teacher described in detail what had to be done. She demonstrated how to measure the pet with the assistance of one of the children who knelt down and pretended to be a dog, while the teacher laid string along from the tip of his nose to the end of his 'tail'! The children all very much enjoyed this session. It kept their enthusiasm and they found it highly entertaining that one of them got to be a dog! The teacher then produced the two pieces of string that she had measured her own dog with at home. The piece that went round his middle was considerably longer than the piece measuring the length! The teacher pointed this out to the children and they discussed this with some amusement.

In Example 1, the activity needed a great deal of preparation, some of which was indirect. In other words, the children were doing a fair bit of bread-and-butter number work, and the IMPACT activity arose out of this as much as anything. Other parts of the preparation were fairly direct in that they led specifically to the IMPACT activity. In Example 2, the only preparation needed was the session on the rug. However, that was crucial in motivating the children and sending them home enthusiastic.

Follow-up Work

The work which follows an IMPACT activity can also be directly related to the activity or it can be more general classwork, connected only indirectly to the actual activity. Once again it is impossible to specify precisely how much follow-up work should be done since this, like the preparation, will depend upon the factors related to age of children, type of activity and so on.

Example 1 Pocket money double-up!
In the case of the pocket money IMPACT activity, the follow-up consisted of two pieces of work. Firstly the children had a long discussion as a whole class, in which they shared their work and explained what they had found out and how they had done all the sums. This session resulted in two main questions: What were the correct sums, since there was considerable discrepancy amongst the children as to the exact figure. How much pocket money did you make in all, since most children had gone on doubling and worked out how much was due on the last day of March, but had not realized that they had to add up the amounts for each day to get a total! Therefore the teacher set the children in groups with two tasks to complete. Firstly, they had to check the sums. All the work done at home was shared here, so that those who had not had a chance to do the IMPACT at home had access to the figures. Secondly, they had to see if they could work out

IMPACT in the Classroom and the Home

how much the total would be *without* adding up all the amounts on each day. (This means finding the binary pattern (see Figure 4.10).

The work at home generated a good discussion, produced some discrepancies which needed checking, and a small amount of new work for some of the children. In addition, it fitted well with their work on large numbers which they were doing as their bread-and-butter maths work.

Example 2 Pet measuring!
This activity generated almost the entire week's maths for the whole class. On the whole, the children worked in pairs and small groups. They compared their pieces of string by lining them up on a base line and seeing which was the longest. They then measured their strings with a variety of non-standard measures – multi-link, beads, crayons, bricks and so on. Some of the children measured their strings using centimetres. All the children participated in recording the information graphically. Some made string graphs, using the actual pieces of string and ordering them. Other children made block graphs using the non-standard measures. A few children made graphs depicting the length of their strings in centimetres.

Further work in groups developed the circumference aspect of the IMPACT activity. The children measured around their own wrists and their heads. They then went on to comparison work with these. Finally, they all drew pictures of their pets and mounted them to make a display.

The IMPACT activity cycle can be regarded as the basic unit of work in delivering the maths curriculum. Because the work is
(a) planned in advance
(b) all related to a central theme
(c) calibrated against attainment targets
it is easy to fit with any system of National Curriculum record keeping.

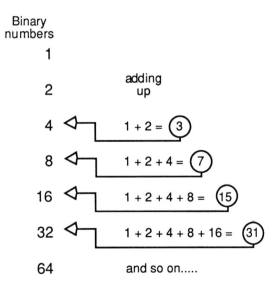

Sharing Maths Cultures

Assessment and Record-keeping

Records

Any system of record-keeping has to fulfil three criteria:
1. It must not only be compatible with the National Curriculum, but also it must enable teachers to produce the information required by Section 22 of the Education Reform Act.
2. It must inform the teacher's planning and on-going allocation of tasks to children.
3. It must be quick and easy to administrate.

Most LEAs are drafting some guidance for teachers on record-keeping, and many schools have developed their own systems.

IMPACT records. Since IMPACT works throughout eighteen authorities from West Glamorgan to East London, and from Humberside to Devon, it has been essential that the systems of keeping records on IMPACT have been adaptable in order that they fit with the local guidelines.

Most teachers on IMPACT make their own chart for keeping an ongoing record. The chart is drawn out by the teacher on a large sheet of sugar paper at the beginning of term. It is mostly drawn out at the same time as the teacher plans her/his work, and it can best be seen as a part of that termly or half-termly planning process. This chart enables the teacher to record not only IMPACT but also all the other maths work as well. The basis of the charge is the block of work surrounding an activity cycle. A specimen record sheet is shown in Figure 4.11.

The teacher writes in the tasks, worksheets and activities that the children do in any one week/fortnight as a part of an activity cycle. She keeps the chart space with her work, so that work which was planned but not done is not entered. Conversely, work which arose spontaneously out of the classroom situation is recorded.

ATs levels		Fred	Joey	Anna	Mary	Matthew	Kim	Mono
2i 8i/ii	Doing Coin sort							
3ii 8ii	Adding up purses							
1ii 3ii 8ii 9ii	IMPACT – How much is your hand?							
3ii	Comparing amounts (subtraction)							
1ii 3ii 8ii 9ii	How much is your foot							

Record sheet (Specimen)

Points to note
1 It is important not to forget the 'incidental' maths which arises as a part of other activities. For example, when a teacher is doing music, the children may be asked to clap their hands ten times. For some children this is an important counting activity.
2 The Statement of Attainment numbers should be added to each activity. This sounds worse than it is since teachers have very quickly familiarized themselves with these. If they are on the chart, it does make it simple to transfer the information to any National Curriculum information form.
3 Many teachers devise some means of indicating on the chart how a child has got on with a particular activity. They usually have some system of indicating whether the teacher thinks that the activity was
   (a) completed satisfactorily
   (b) only partly understood
   (c) in advance of the child.

It is even more helpful if these remarks are related to particular statements of attainment.

Parental input. There is little point in involving the parents on a weekly basis and not making some provision to allow for the parent's comments to contribute to the child's records. The work done by the Records of Achievement Working Party (RANSC) shows clearly that an input by both parents and the pupils is of value in making the record more meaningful.[39]

Within the structure of the activity cycle itself there is a place for an input by the child and the parent when they comment on how the home maths part went. We shall be discussing the various ways of obtaining, encouraging and structuring these comments later in the chapter. However this is done, these comments form an intrinsic part of the activity cycle and they therefore have a natural place in the records.

We have again resisted the temptation on IMPACT to develop a standard procedure or format for these records, but teachers have drafted their own means which are compatible with their personal recording systems. There are three common procedures:
1 The teacher adds a column to the existing record chart. Since this is large, and personal in the sense that the teacher her/himself draws it up, it is a simple matter to add a column or a line. The advantage of this method is that the parents' and children's input is a part of the main class record. The disadvantage is that the record of the comments has to be very brief and may lack detail.
2 The teacher keeps a separate record of all the parents' and children's comments. This enables this record to be more detailed than the above. It is often kept in the register for ease of access. The advantage of this system is that the complexity and variety of the comments can be preserved. The disadvantage is that the two records are separate and therefore have to be linked. (See Figure 4.12 for examples of records.)
3 The children's and parents' comments are themselves preserved in a folder for each child. This folder makes an on-going record for each child in

IMPACT (7) COMMENTS RESPONSES w/e 25/11/88.

	Name	Stat.1 Out	Stat.2 Returned	Subject OWN CHOICE BOX (1) Comments
4th years.	Nikki.	✓	✓	Nikki understood this very well but we went a little wrong - Sister. (Aged 13). Katy saved the day - Ch.
	Catherine A.	✓	✓	Understood activity very well - P.
	Robert.	✓	✓	No comments
	Dale.	✓	.	
	James.		/	Absent.
	Roxane.	✓	✓	Picked possibly the hardest one to make & found it too difficult - P.
	Christopher.	✓	✓	C. understood the activity v. well - P.
	Michael.	✓	✓	No comments.
	Stuart.	✓	✓	No comments.
	Catherine S.	✓	✓	No comments.
	Charlotte.	✓	✓	No comments
3rd years.	Colin.	✓	✓	No comments
	Sam.	✓		
	Dawn.	✓	✓	We had to find a way, without compasses or protractor, to find the apex of an equil: △. - P.
	Paul.	✓	✓	He understands the activity well enough to do it by himself - P.
	Brian.	✓	✓	B. had no problems at all with activity - P.
	Victoria.	✓	✓	V. understood the activity & asked for no help - P.
	Gareth	✓		
	William.	✓	/	Absent
	Anya-Kristina	✓	✓	She understood it fully, wanted no help, worked with a friend. - P.
	Marc.	✓		No comments
	Rebecca.	✓	/	Absent.

IMPACT in the Classroom and the Home

IMPACT (7). RESPONSES. w/e 25/11/88.

Name	Activity	Degrees of Difficulty	Amount of Help	Amount Learned	Follow Up	Liked	Activity Was	I Learned	Time
Nikki Kirk.	Quite	Right	Little	Nothing	Enough	OK	Right	Lot big	50
Catherine Addison	Quite	Easy	No.	Little	Enough	OK	Easy	Little	30
Robert Gibbs.	Very	Right	No	Lot	Enough	Very	Right	Lot	45
James Pickard.									
Roxane Kedge.	OK	Hard	Lot	Little	Little	OK	Right	Nothing	60
Chris. Hayward.	Quite	Right	No.	Little	Enough	Very	Easy	Lot.	25
Michael Goss.	Quite		No.	Little	Enough	Very	Easy	Nothing	10.
Stuart Bryant.	Very	Right	No	Lot	Enough	Very	Right	Lot	45
Catherine Spencer.	OK	Right	No	Little	Enough	Very	Right	Little	35
Charlotte Cross.	OK	Right	No	Little	Enough	Very	Right	Little	10
Colin Townsend.	Quite	Right		Little	Enough	Very	Right	Little	30
Sam Burden.									
Dawn Crichton.	Very	Right	Little	Little	Enough	Very	Right	Little	40
Paul Smith.	Very	Right	No	Little	Enough	Very	Right	Little	20
Brian Darst.	Quite	Bit	No	Little	Enough	Not	Easy	Little	10
Victoria Evans.	Very	Right	No	Lot	Little	Very	Right	Little	45
Gail Whiting.									
Gareth Gummer.									
William Dew.									
Anya-K. Kroeger.	Very	Right	No	Little	Enough	Very	Right	Little	20
Marc Walker.	OK	Right	Little	Little	Lot	OK	Hard	Little	30
Rebecca Blythe.									

which all their personal records are kept and which passes with the child up the school. Once again, this method preserves the detail of comment, but it lacks the advantage of an intrinsic connection with the other class records.

The input of parental comment into the school and class records raises a number of questions. Many teachers feel that the comments about certain types of activities are not accurate. For example, many parents seem to write that an activity was too easy, unless it was actually too hard. There are then reservations about transcribing these comments into official records. The questions and implications raised here are discussed later on in this chapter under the heading of 'parental responses'.

Assessment

With the National Curriculum it is basically no longer possible to separate record-keeping from assessment. As we observed earlier, the whole curriculum is now described in ten levels of attainment. This means that when we record what a child has done, we are automatically making a partial assessment since we are simultaneously recording at what level he is operating. This has implications for the type of records we keep and the way that we think about the parents' and children's responses.

Teachers will be routinely involved in both the administration of Standard Assessment Tasks and continuous teacher assessments. However these change and are adapted over the next few years, two things are more or less inevitable.

1 Teachers will be looking at children with reference to attainment target levels, and they will be formally recording at what level they think that the child is operating. There will be a tendency, simply because of the size of the curriculum, the numbers of statements of attainment and their detail, to focus solely upon children's progress in relation to these. The monitoring and formal recording of qualities such as attitudes, perseverence, concentration and so on, may take second place for the majority of children. Similarly, it will be tempting to focus upon the immediate performance of skills, rather than the child's ability to transfer these skills or use them in a variety of alternative situations.

2 Judgements made about which tasks, worksheets and activities the children will do in class will be more crucial than ever, since part of the assessment process consists in recording at what levels the children are operating in their routine work. This means that, in theory at least, a teacher could be affecting a child's assessment profile directly by a mismatch between task and child. The HMI reports indicate that, for the least able and most able children, there is often such a mismatch.[10]

Parental input into assessment. Inviting a parental input can enable us to address some of these issues. On IMPACT, in several schools, once or twice a year, the parents are asked to complete an assessment sheet about their child. In some schools this is a sheet that the parents take home and fill out, after discussion with the teacher. In some schools, this is filled out more like the ILEA Primary Language Records (PLRs), where parents and teachers discuss the sheet together and fill it out as a joint exercise. Parents are asked to make their own judgements as to whether the child has acquired certain skills, and to what extent they are comfortable with their application. These judgements are based upon the work sent home on a weekly basis, but they will also be affected by the parent's own dealings with the child in everyday routines. Do they find the child can tell the time? Can they work out change? Do they recognize coins? IMPACT activities may focus attention on some of these skills but everyday life will provide its own information as well.

Taking into account parents' judgements as to their children's assessment profile will have the following advantages:

1 At the very least, it provides a check on the teacher's judgements. In most cases, there is substantially agreement, and it is perhaps just a few details which the teacher will feel need to be checked. In a few cases, there may be major discrepancies between the teacher's judgements and the parent's and this will need to be the focus of attention, both in terms of a dialogue between teacher and parent, and in terms of some further assessment of the child.
2 It provides a check on the teacher's selection of task for the child. If the parent says that a child has not acquired a skill, such as telling the time, which the teacher had assumed, this will affect the teacher's choice of work for that child. Parents' comments here can be very illuminating since there is substantial evidence that teachers consistently tend to over-estimate confident, quiet, mature children, and under-estimate immature, unconfident, noisy or disruptive children.[41]
3 Parents do not see their children – on the whole – in a classroom situation. They will see their children applying skills in a totally different context. Their judgements as to whether the child can or can't do something are of great value in enabling us to broaden the basis upon which our assessments are made.
4 Parents will instinctively discuss their children in terms of their qualities, such as perseverence, temper, concentration and so on. These will form an explicit aspect of the agenda when discussing parental judgements on assessment.

Once again, there are many implications to inviting a parental response in this way. We discuss these in the next section.

Listening to Parents

We have spoken a great deal about the importance of using parental and children's responses in records and assessment. This section is concerned with
- ways of encouraging and obtaining a response

Sharing Maths Cultures

- different types of response
- the implications of what is said
- patterns of parental contact.

The IMPACT Project has been at the business end of work in this area. What we sometimes regard as normal and accepted practice can often be highly contentious. We address some of the issues discussed in IMPACT in-service sessions.

Ways of Encouraging and Obtaining a Response

In keeping with IMPACT practice, we have not insisted upon the precise format which schools should use in order to obtain comments from the parents and children. However, we have, as far as possible, tried to insist that there is some structured means of obtaining a written response on a weekly basis. We are aware that some teachers and schools have argued that they prefer to encourage parents to come in and chat about how the activity went at home, and that they feel that a written comment is too formal, or even intimidating. However, our experience on IMPACT leads us to believe strongly that:

a) Having a written response form or diary does not preclude, or even discourage, a verbal response as well.

b) In practice, there is a tendency for parents, especially of older children, to find it difficult to make the time to come into school. On a weekly basis, simply to make a comment on an activity, most parents will not bother unless something went wrong. Thus, an on-going and continuous commentary is not obtained.

c) If all, or most, of the parents did come in for a chat, would the teacher have time to cope? If the answer to this is no, then we are not really serious about inviting them to do so and the parents will recognize this, unconsciously if not consciously.

d) Providing a means by which parents and children comment in writing about how the activity went gives a clear message to parents that we do value their comments. Written responses have a status denied to verbal comments in our society. We may deplore that this is so but we cannot ignore it.

e) Asking a parent to come into school and talk about what happened at home with an activity may be just as intimidating as asking them to tick boxes on a comment sheet. Teachers may perceive a 'chat' as being informal, but it is by no means always seen that way by parents owing to a perceived asymmetry in the relationship. 'Perhaps you would like to drop in and tell me about it', when said by a headmaster to a pupil is not necessarily as informal for the pupil as it is for the adult.

f) It is not necessary that a parent needs to be able to read or write in order to fill out a comment sheet. The experience of many years' successful shared reading initiatives in different parts of the country has given us a variety of mechanisms for obtaining feedback without relying upon the reading or writing of the parent or child.[42]

g) It is easy to translate the comment sheets into the first language of the home, where this is itself a written language. On IMPACT, we have comment sheets in Urdu, Bengali, Greek, Welsh and twelve other ethnic minority languages.

Designing a response form. We have always worked with the teachers in any area to design a system of comment sheets or diaries which those teachers felt was appropriate for their schools. These have usually had the same basic format but have varied considerably in the number, type and wording of the questions.

The areas where we traditionally invite comments from parents include:
- How easy or hard the activity was.
- How enjoyable it was.
- Whether they felt that the child learned anything.
- If they think that the child needs more practice or help with this aspect of maths.
- How long it lasted.
- Who did the activity with the child.
- How much help was needed.

The areas where we traditionally invite comments from children include:
- How easy or hard they found the activity.
- How much they enjoyed it.
- Who helped them.

We always include a space in which parents and children can write about what they did, if anything interesting or unusual or exciting happened. Parents also commonly use this space to express concern either about something they feel their child is having problems with, or an aspect of the curriculum which is puzzling or worrying them.

Layout and type of form. Comment forms divide into two main types: diaries and sheets.

a) Diaries
These also divide into two types.
Type 1
There are those which serve a dual function of enabling the parent to respond to both the shared reading at home and the IMPACT activities.

This type of diary uses the same method for IMPACT as for the comments on the reading. Basically, the parents write comments and the teachers write back. In this example, the IMPACT comments are structured by some guidelines in the front of the diary but it is up to individual parents whether they take any notice of these, or whether they decide to write freely as they do on the shared reading.

Advantages
1 This type of diary does allow for a parent-structured response. Of course, it is always true that we prejudge the answers by the questions we choose

Sharing Maths Cultures

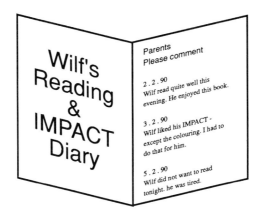

IMPACT and Shared Reading Diary

to ask, and the questions address 'teacher' preoccupations. In a diary in which parents, children and teachers write freely, it is possible for the parents — or indeed the children — to set the agenda.

2 Teachers and parents do have an on-going dialogue through these diaries. This can be very helpful to both and can make a great deal of difference to the nature and quality of the information which passes between home and school.

Disadvantages

1 Some parents find a blank sheet very intimidating. The unstructured approach does nothing to help those who are not accustomed to communicate in writing.
2 The diaries are not good for those who have trouble either reading or writing.
3 The demands on the teachers are heavy since they may be responding in writing on a weekly basis to thirty children.

Type 2

These are diaries intended solely for comments on IMPACT and they tend to reflect a more structured approach.

In the example given, the diaries are constructed so that it is not necessary to be able to read. Once the original categories have been explained, the parents can respond simply by a series of ticks across the faces. Space is also left for any written comments the parent or child wish to make. The whole diary is just a few pages of A4 paper folded and stapled. some schools add a card cover, others leave the children to decorate the outside sheet of paper.

Advantages

1 The diaries are economical, in that a whole page serves four weeks of IMPACT. Thus, the children have one diary for the whole year.
2 The diary keeps an on-going and highly visible record. This means that parents and teachers can see at a glance several weeks' comments. If a

IMPACT in the Classroom and the Home

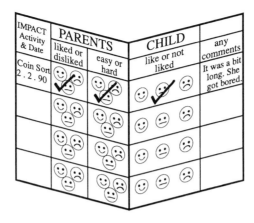

IMPACT Diary

parent is continuously writing the same thing, this will be immediately apparent.

3 The diaries last a whole year and can therefore be used as an individual child record. They can be kept as a part of the child's folder.

4 Our experience on IMPACT is that these diaries are extremely unthreatening and that they produce a very high response from parents. Many parents also fill out the 'comment' slot. It is as if the faces to be ticked do not quite enable the parent to say what they want to say about how the activity went, and so they find themselves writing a bit to explain further.

Disadvantages

1 The highly structured aspect of the diaries means that the parents' comments have to fit neatly into the pre-defined categories provided. This limits the range of response, and may even preclude certain honest parent-directed comments.

2 The scope for writing is limited by space, and this means that to some extent we are conveying to parents a sense that their written comments are less important than the face-ticking exercise.

* Stop Press. These diaries may now be purchased in bulk from IMPACT Central Office (see Appendix).

b) Sheets

There are as many different types of sheet as there are LEA on IMPACT, and some more besides. They are all laid out in the same manner. An A4 sheet is divided down the middle as shown in the example, and the left hand side is the parents' half, the right is the children's.

The questions vary from school to school, from teacher to teacher. But we do feel that it is important that parents' opinions are sought on certain aspects, and so the advice we give will suggest that questions are included on the enjoyment, on the level (easy/hard), and on the amount of help needed and given.

Sharing Maths Cultures

IMPACT   IMPACT   IMPACT   IMPACT   IMPACT

CODE **Hd 3**

Activity name **Number Detective**   no. **2**

1. Write your name **Aneetha**

2. Did you like this week's activity?

   not much   it was O.K.   I liked it ✓

3. Did you think it was

   too hard   too easy   just right ✓

4. If you want, write about what you did. When I first brought the game Home, I had to tell my mum how to play the game. At first when she saw it she didn't know what it was all about. Then I told her how it works and she said it will be easy. And we had good fun playing the game

Advantages
1 This system is very economical because the comment sheets can be ordered and printed in advance. Then the activity which the children are to do can be copied on to the back. This means that one sheet of paper per child serves the dual purpose of conveying what is to be done, and inviting comment upon how it went.
2 The teacher knows precisely which activities each set of comments applies to because the activity itself is on the back of the comment sheets. Even when children bring them back a little haphazardly and on the wrong day or, worse still, the wrong week, it remains easy to establish which comment pertains to which activity.
3 The sheets combine many of the best aspects of both types of diary. They can make it easy for parents to comment even without reading or writing, and there is plenty of space for free comment. This means that parents can, and frequently do, write a whole page describing what happened or expressing a particular activity.
4 It is easy to translate the comment sheets into as many ethnic minority languages as necessary. On IMPACT, we have already done this and specimen comment sheets may be obtained from us. (See appendix.)

Disadvantages
1 It is necessary for the teacher to be efficient about transferring the information contained on all the comment sheets as fast as possible into his or her records. Otherwise he or she runs the risk of drowning in paper, since in our experience these sheets seem to multiply and get disordered without any human assistance!
2 It is difficult to persuade some parents that the on-going comment helps. The IMPACT comment sheets look like 'just another form' in a bureaucratic age. The teacher has to make efforts to convince the parents, more by actions than words, that their comments are read and valued.

Different Types of Response

The difference in format and layout of the means of obtaining a response will clearly affect the nature and type of responses obtained. With this in mind, it is important that the teachers in any one school do give careful consideration to the sort of comment sheet or diary that they want to use. On IMPACT, rather than providing one standard model, which would undoubtedly suit some but not others, we have been content to supply a range of examples of accepted practice. Figures 4.16–20 show some of these.

We have found that parents tend to respond in certain ways to the IMPACT activities, especially at the start of the initiative.

Number of responses. We cannot say that schools in a particular catchment area are more — or less — likely to get a certain percentage of responses in terms of just how many parents actually share the activity with their child. The numbers of parents doing so does vary from school to school and even

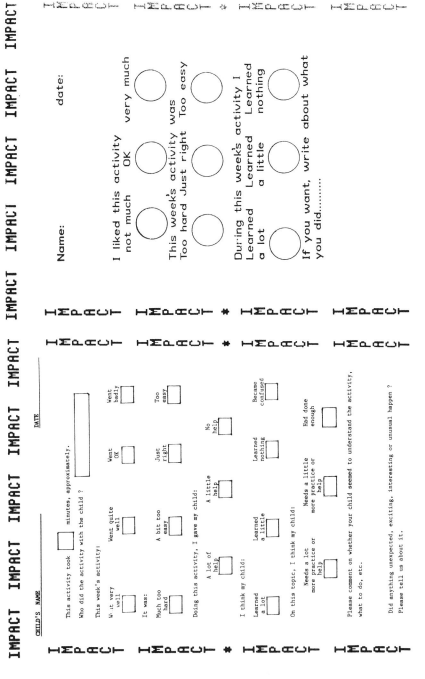

IMPACT in the Classroom and the Home

Name: **Steven**

Did you enjoy the activity?

Why? I do enjoy doing things like this with Steven, because he looks pleased he thinks he is teaching me.

If you want, write or draw about what you did.

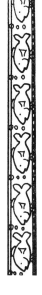

Helper's Sheet   Date

Name of Activity **Spans**

This activity took **18** minutes

Who helped the child?

This weeks activity went: very well / had problems ✓ / no problems

If you had a problem, what was it?

I worked with my child: all the time / Most of the time ✓ / Some of the time

Please comment on whether your child understood the activity and on how it was completed.

Yes he understood very well

IMPACT   IMPACT   IMPACT   IMPACT   IMPACT   IMPACT

Adult's Questionnaire

Date_____   Time taken _____

Type of Activity _____   Eg. weighing, measuring, etc

Did the child undertake this activity   willingly/after mild persuasion/
                                            not willing to do this activity

Did you give the child: no help/a little help/ a lot of help

If a similar activity was prepared for next week, do you think he/she would;

complete it with no problems/need a little supervision/
need your help each step of the way.

Which of the following statements are true?

He/she was confused at the beginning & still confused at the end.
A little unsure at the beginning but slowly gained confidence.
Was confident at the beginning but slowly became bored.
Was enthusiastic throughout activity.

Any further comments or queries about this activity.....

IMPACT in the Classroom and the Home

IMPACT IMPACT IMPACT IMPACT IMPACT IMPACT

I
M
P
A
C
T

Name _____ Date _____

Activity: _____

How long did this activity take? _____

Did you like this activity? _____

Why? _____

During this week's activity I learned:

    A lot   A little

Can you write or draw about what you did.

If you need help filling in this form ask someone to help you.

IMPACT IMPACT IMPACT IMPACT IMPACT IMPACT

IMPACT IMPACT IMPACT IMPACT IMPACT IMPACT

I
M
P
A
C
T

CHILD'S NAME _____ DATE _____

NAME OF ACTIVITY _____

This activity took ⬡ minutes.

Who did the activity with the child? _____

This week's activity:

  Went very well   Went OK   Had Problems
    ⬡   ⬡   ⬡

If you had a problem what was it? _____

Doing this activity I gave:

  A lot of help   A little help   No help
    ⬡   ⬡   ⬡

What do you think the child gained from this activity? _____

Did anything unusual, exciting, interesting, unexpected happen? Please tell me about it.

_____
_____
_____

IMPACT IMPACT IMPACT IMPACT IMPACT IMPACT

Sharing Maths Cultures

Maths is Fun.

I.M.P.A.C.T.
Diary

Name _____

IMPACT in the Classroom and the Home

Date	Activity	How did it go?	Was it?	Did you give?	I think my child	On this topic I think my child	Other comments you would like to make
		Very well ○ Quite well ○ O.K. ○ Poorly ○	Much too hard ○ A bit too hard ○ Just right ○ Too easy ○	A lot of help ○ A little help ○ No help ○	Learned a lot ○ Learned a little ○ Learned nothing ○ Became confused ○	Needs a lot more practice ○ Needs a little practice or help ○ Has done enough ○	
		Very well ○ Quite well ○ O.K. ○ Poorly ○	Much too hard ○ A bit too hard ○ Just right ○ Too easy ○	A lot of help ○ A little help ○ No help ○	Learned a lot ○ Learned a little ○ Learned nothing ○ Became confused ○	Needs a lot more practice ○ Needs a little practice or help ○ Has done enough ○	
		Very well ○ Quite well ○ O.K. ○ Poorly ○	Much too hard ○ A bit too hard ○ Just right ○ Too easy ○	A lot of help ○ A little help ○ No help ○	Learned a lot ○ Learned a little ○ Learned nothing ○ Became confused ○	Needs a lot more practice ○ Needs a little practice or help ○ Has done enough ○	

from class to class. (See Chapter 1.) However, the social class of the parents does not seem to be a factor. The variation in response has to do with a number of factors, of which 'teacher enthusiasm' is certainly one of the most important. For example, within one school in a socially deprived inner city area, we have a situation where three teachers are finding that 80–90 per cent of the children share their IMPACT activity, and one teacher in the school finds that only 30 per cent of the children do so.

Other relevant factors include:
- the extent to which the IMPACT activities are perceived as part of the classwork by the children, rather than simply an optional extra;
- the age of the children. Overall, top juniors tend to respond less than younger children, at least initially, and the nursery and middle/top infants are always especially keen;
- the amount and quality of the follow-up work in class.

Number of comments. There does seem to be a correlation between the social class of the catchment area of the school and the number of comment sheets which parents fill out. Although, as stated above, social class does not appear to be a factor in determining how many parents actually share the activity at home, it is a factor in predicting whether or not the teacher is likely to get many comments back. It is also a factor in determining what these comments are likely to consist of.

What parents say. We have found that overall there is a strong tendency for teachers and parents to disagree about whether a particular activity was too easy or not. Parents tend to say that an activity was too easy where the teacher is convinced that it was just right or even a bit too hard. There are a number of ideas about why this might be.
- Many teachers feel that parents say that an activity was too easy unless it was actually too hard.
- Some parents feel that teachers can under-estimate what children are capable of.
- Children operating in the security of their own home may be able to do things or perform in ways which are unlike how they perform at school.

We have also found that children and parents frequently disagree as to whether an activity was enjoyable or not. They also do not necessarily agree about the level of an activity, its easiness or hardness.

Some parents are inclined to tick the 'needs more help or practice' box on the basis that more help cannot do any harm and might do some good. The idea that practising a skill already acquired might be damaging to the child's attitude, motivation, perseverance or even to the skill itself, is a novel one for most parents.

The Implications of What is Said

The in-service sessions in which we are designing comment sheets with teachers prove to be some of the most interesting and lively sessions in our experience. The discussion which takes place as we try to work out which questions to ask parents and how to phrase them, raises all sorts of issues concerning the way we as teachers see our role, and the boundaries of professionalism. From the experience of many such sessions we have recognized the recurring concerns which seem to be common to all teachers, whether they are infant, nursery or junior, in Hull or Exeter, in the inner city or in rural villages.

The question of level. There are many teachers who feel that to ask whether the activity was too hard or too easy is a bad idea. They argue that parents cannot tell if an activity is too easy or not because they are unable to see the way an activity can be helping a child, or to recognize its mathematical value. This is not because the parents are not as intelligent as the teachers but simply a result of not having had the same professional training.

However, other teachers disagree. They feel that if the question is not asked, then what can be useful information is not obtained. For example, the activity sent may be too easy or too hard. A teacher may be underestimating or overestimating a child, or may not realize what the child can achieve when in a one-to-one situation at home.

Our view on the Project is that it is a question of the sort of information we think we are both asking for and getting. We are not asking the parents for a definitive answer to the question of level. We are not proposing to abandon our own sense of the level of a particular activity *vis-à-vis* a specific child. What we are doing is setting up a situation where there is a dialogue between the two — possibly opposing or contradictory — perceptions of the parents and the teachers. Teachers *cannot* know how a child will perform at home. Parents *do not*, on the whole, share the teachers' background knowledge and assumptions which they have acquired through professional training and experience. But out of the dialogue which we set up between these two, can come a much richer picture of the child's competence and potential.

Therefore, it is important that we do ask such questions and that we are both clear and honest about how we are reading the answers.

If it doesn't hurt, it's not doing you any good. There is a tendency for parents — and some teachers — to think of maths like medicine. If it doesn't hurt, if it tastes nice, then it is clearly not doing you any good. Most parents unfortunately have bad memories of maths at school, when it was a byword for boredom and bewilderment. When the children are clearly enjoying what they are doing, the parent is cast into doubt as to whether they are actually learning anything! This affects their expressed views on the difficulty of the activity. It can form the subject of a discussion between the parents and the teachers, especially if the parents are invited to see how the IMPACT activities are followed up in practice, and where they lead to in terms of mathematical understanding.

There's no point in all this. Another issue which arises out of the dialogue between parents and teachers generated by the response forms is the question of the mathematical *point* of many practical, problem-solving or investigational activities. It is often not only the parents who are not too clear about this. Sometimes a teacher will be doing an activity with the children without being fully conversant with its rationale in terms of the children's mathematical development.

For example, many teachers do a series of tasks and worksheets with children using 'Tangrams'.

Some children can have real fun with these and teachers are convinced by the commercial schemes or Teachers' Centre booklets which produce such activity outlines. But where do they belong within the curriculum? What particular aspect of mathematics are they assisting or illuminating? Unless a teacher — or a parent — is particularly well qualified in mathematics or psychology, it may be difficult to answer such questions.

The National Curriculum has given us the Programmes of Study, and it is within these that teachers now plan their work. They will be able to see from their plans where the tasks, worksheets and activities fit into the overall scheme of work. This may involve the teacher in questioning why certain activities are used at particular stages. However, having engaged in this process, a teacher is better able to discuss with parents the purpose of particular activities, and whether they are worthwhile.

What about 'right' answers? Genuine differences of opinion between teachers and parents on the whole approach to the teaching of mathematics do occur. In the post-Cockcroft years, many teachers have become convinced of the place of practical and open-ended, problem-solving activities in developing children's mathematical understanding. However, many parents still see maths as a hierarchy of skills and procedures, most of which are best rote-learned at a young age. With the advent of the National Curriculum and its emphasis upon skills, a new set of demands and their rationales are placed in the arena.

On IMPACT, we have not found a set of easy answers to these questions. But we have found that children do much better when parents and teachers talk about these things. If the school adopts one approach to maths and the parent believes in another, the child will at best get no support and at worst get contradictory advice. 'Try it out, investigate it and see if you can understand why it works . . .' in the classroom, and 'Never mind about why, just copy how I do it and you can't go far wrong . . .' at home.

Uncomfortable conversations. However, conversations between parents and teachers on these issues are not always easy. Teachers can feel very threatened by the suggestion that children are 'playing about' and not learning. It can be difficult if not impossible to demonstrate exactly how the approach taken is superior to the more 'traditional' way of teaching maths, especially if the parent was one of those who learned 'successfully' under the old system.

Some parents have complained that despite what they say about home/school liaison, teachers do not in fact listen to parents' comments. 'They have already made up their minds how they are going to teach maths and any comments made by parents simply fall on deaf ears.' There is little point in writing responses, such parents feel, when nothing they write is going to change anything.

IMPACT *in the Classroom and the Home*

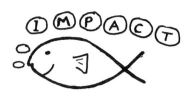

Tangrams

This is an old Chinese puzzle. It is useful to demonstrate that the area of all the shapes is the same, although some seem much larger.

1. Cut out the tangram pieces carefully.

2. Rearrange them to make the following shapes and record the arrangement by sketching. Use all the pieces.
a) rectangle   b) triangle

3. Use the pieces to make animals, letters, numbers, transport shapes or anything else you like. Draw each shape to show how you fitted the pieces together.

4. Colour in your work to be displayed.

Extension:
a)   Try a parallelogram

b)   Make a trapezium.

Midwives and mothers. It can be helpful to use an analogy here. A midwife is a professional person. Mothers-to-be are 'amateurs'. But it is a foolish midwife who doesn't listen to the mother. Similarly, it is a foolhardy woman who doesn't accept that the midwife's experience and professional knowledge is essential. A 'nice' birth depends upon mothers and midwives both talking and listening. At a certain point in the birth of each of my own six children, I relied totally upon the expert advice of the midwife. At other points she relied heavily upon my description of my state of labour.

A wise teacher realizes that parents have another perspective, and will listen to what the parents have to say. Teachers do not (unfortunately!) possess any God-given knowledge on the best way to teach maths. They have some theories, and some experience, and they will be aware of the non-obvious problems which may not be immediately apparent to parents. But sometimes their approach is not succeeding as they had hoped it might. Sometimes minor adjustments are needed, or a little adaptation to particular circumstances. Parents' comments can alert teachers to possible causes of trouble or difficulty before they become serious or necessitate drastic action.

Thoughtful parents also realize that teachers do not go to immense trouble to give children practical tasks to do, and to generate problem-solving situations, simply in order to make life more complicated. Professional experience demonstrates the dangers of allowing children to engage in formal written procedures or mental algorithms without also developing an understanding of what they are doing. We also have ample evidence that if children do not see the point of a particular procedure, they are unlikely to retain it or be able to transfer it to real life situations. And all parents will agree that a skill that can only be used in a maths book in the classroom is not of much value.

Patterns of Parental Contact

We have talked a lot about developing an honest and open dialogue between parents and teachers. Of course, such a thing is only possible if time is allowed for the conversations to take place. And many teachers feel that they are already committed to working twenty-five hours a day in order to deliver the National Curriculum. How is the time to be found to talk to parents?

On IMPACT teachers have developed a variety of ways of giving each other time in order to allow for conversations with parents. These include:

a) Story-round

Once or twice a week, the teachers in the school, or in one part of the school (infants, juniors, or even a couple of year groups in a large school), will agree to run a 'story round'. Instead of reading a story to their own class at the end of the day, all the teachers except one say in advance which story they are going to read to anyone interested. The children are allowed to choose which classroom to go to and which story to listen to. The teacher who is 'freed' by this process, then has time to have a cup of tea and a chat with any parents who want to come up to school ten minutes early and talk about IMPACT.

The children have an opportunity to choose a story that they fancy hearing, and it is quite surprising how they exercise this privilege. We have seen top juniors listening to *Not Now Bernard* and middle infants engrossed in *The Hobbit*. The teachers enjoy the chance to read a story they like to a larger — or a smaller — group of children of mixed age. And one teacher gets a chance to talk to parents.

b) Class teas

Each class takes it in turns to have a class tea on a particular day of the week. For example, on Wednesdays the parents know that they can come into school half an hour early and stand or sit around in the hall and drink a cup of tea with the teacher, headteacher and any other available staff. The children help by serving cakes, usually brought in by parents, but also made by some of the children in the class.

Each class has a turn at providing the tea and cakes only once every five or six weeks, so the burden on the parents in supplying cakes and biscuits is not that great. Any parent can come, not just the parents in the class running the tea. The occasion runs across the end of school, so that, if school ends at 3.30 p.m., class tea starts at about 3.10 p.m. and goes on till 4.00 p.m. This means that all the teachers, even those whose class is not involved can join in after the children have been dismissed. This occasion provides a friendly and relaxed way of enabling parents and teachers to talk informally on a regular basis.

c) Problem-solving mornings

Once a week, one teacher (who can be the headteacher) organizes and runs a 'Problem-solving event'. This consists of a chance for parents and children to come into the hall and work on a particular problem in groups or pairs. For example, they may be asked to build a shelter big enough for a child using only five old newspapers, string and Sellotape. Or they could be required to design and make a parachute out of a black dustbin bag, string and a paper cup, which will carry an egg (unboiled!) safely down a distance of ten feet.

The teacher organizing this event sets it up for three or four classes. The other teachers are also in the hall, not so much to help the children, but to talk to any parents who want a chat.

The beauty of this arrangement is that the discussions take place within a real mathematical event. It does mean that conversations tend to be between the sellotape-tearing and the newspaper-folding, but they are none the less illuminating or honest for that.

d) 'Class swaps'

More conventional ways of enabling teachers to have time to talk to the parents in their class involve agreeing to read each other's story, or take the register, once a week. Thus, Mr Bloggs takes Miss Fung's class as well as his own for story on Wednesday, and Miss Fung does the same for Mr Bloggs on the following Wednesday and so on. Every other Wednesday the parents in one or other class have half an hour at the end of school in which they can talk to their child's teacher about IMPACT.

Initial Parents' Meeting on IMPACT

In order to start the project off in any school or class, it is essential to try to speak to as many of the parents about IMPACT as possible. We have found that there is no substitute for eye-to-eye contact and face-to-face conversation. A letter simply does not carry the same sense of excitement, enthusiasm or commitment.

We have found that it is possible to get between 70 and 80 per cent of the parents into the school for a meeting even in schools which have traditionally had no success at all at persuading parents to attend. In some schools, we have found that 100 per cent of the parents have turned out.

The method we adopt is as follows:
- We send a letter home (*in translation if necessary*) which outlines what we are about to start with IMPACT, and says that we really would like to talk to parents for just ten minutes about this.
- The letter offers parents four times of day, on a particular date, at which they can come into school to a *short* meeting. It has a tear-off slip on the bottom which parents have to fill out to let the school know which meeting they will be attending.
- The letter emphasizes that small children are welcome and that any children can be brought to the evening meeting.
- On the day of the meetings, a teacher acts as 'sheep-dog', and stands out in the play-ground shepherding parents into the meeting.... 'Had you remembered our IMPACT meeting?... It's only ten minutes.... There's a nice cup of tea....'
- The meetings are only ten minutes, and serve simply to put across the following points:

    a) From now on we are going to be involving you in your children's learning of maths.

    b) You do not need to know any maths.... We want them to explain to you what they have to do each week. We think that it is really helpful to the child to have to explain some maths that they have learned at school to someone at home. As they do this, they make that knowledge their own.

    c) IMPACT activities are different from traditional homework in that they are designed to be shared. The child is *not* intended to do the activity on their own.

    d) The maths activity they bring home will have been specially designed

IMPACT in the Classroom and the Home

Dear Parents

We are starting a new scheme, IMPACT, to involve parents in their children's learning of maths. This will be running in your child's class and we would very much like to talk to you about it.

We shall be holding 2/3/4 short meetings on _____.
Please choose the time which is most convenient. You need only come to ONE meeting. Small children are very welcome and there will be some toys for them to play with.

If it is impossible to come to any of these meetings, could you let us know so that we can arrange another appointment.

<u>Please return to class teacher</u>

I am able to attend the IMPACT meeting at _____

Signed _____

Child's name ....

by the class teacher to fit exactly with the maths the children are doing in class, week by week.

e) The results of the work at home will be used back in the classroom the following week. All the children will share any work done at home.

f) Don't worry if you don't manage to do your IMPACT one week. We aim at most parents and children doing it most weeks. You can always 'drop back in' if you have 'dropped out'!

g) This assistance that you give your child is of immense value. We are not asking parents to be teachers but we do know that where parents and teachers cooperate, the children's learning benefits greatly.

Points to note. We have found that where such initial meetings have not been held, this has adversely affected the numbers of parents and children taking part. We have also found that it reduces the numbers of parents attending such meetings if a 'tear-off' slip is not included on the letter. Translations of the letter of invitation into the main ethnic minority languages including Welsh can be obtained from the IMPACT Head Office (see back of book). If you have a number of bilingual parents, it may help to explore the possibility of one of the parents — or children — acting as a translator at the meetings.

Follow-up Parent Metings on IMPACT

It is also necessary to have regular follow-up parent meetings on IMPACT in order to monitor how the activities are going. The frequency of these will need to depend upon the school and classes involved. Factors which affect how often these meetings occur include:

- How often teachers and parents meet for informal chats.
- How many of the parents find their way in for these informal chats.
- Whether the school has implemented any of the suggestions given above to enable parents to come in on a regular basis (e.g. Class teas, story round, etc.).

The norm on IMPACT is that such meetings happen about twice or three times a year.

The format of these follow-up meetings is important. We have found that it is possible for one or two parents, especially if they are critical of IMPACT or the school's approach to maths, to dominate such follow-up meetings. This makes the whole thing feel very negative, and it also makes it difficult for other, more nervous parents to have their say. We have therefore adopted a particular method of organizing and running the IMPACT follow-up meetings which enables every parent to have an input whilst retaining the possibility of both positive and negative comments and feedback.

Form of meeting. The seats are put out before the parents meeting. They are arranged in twos and threes. At the start of the meeting, after the parents

have been welcomed, the teacher asks the parents to work together in twos and threes. Each group is supplied with a piece of paper and a pencil. They are asked to make an agreed list of:
- Three positive things that they have found about participating in shared maths activities.
- Three things about IMPACT which could be improved or which give cause for concern.

The groups are allowed ten to fifteen minutes to complete their list of points.

The groups are then amalgamated into larger groups of six. Each large group now has ten more minutes to agree a combined joint list of three points either way. This will involve considerable discussion and negotiation. Finally, one of the teachers agrees to act as scribe, and there is a general feedback session in which the groups report in turn. As they are mentioned, the points for and against are written up on a board so that everyone can see them.

Advantages

There are many positive aspects to this form of meeting organization.

1. No one individual, no matter how loud, forceful, opinionated or belligerant, can dominate the meeting. The most such a person can do is to argue within their threesome, and again within their larger group, so that some of their points are on the agenda.
2. Everyone has a chance to say what they feel and to see whether it is a perception shared by others or unique to themselves. Shy or quiet parents are not excluded as they will talk in their groups of three.
3. A good part of the discussion, and maybe argument, takes place parent to parent within the groups, rather than teacher to parent. This gives the whole meeting a much less confrontational feel.
4. At the end of the day we get three *positive* points as well as three negative ones. This prevents the meeting degenerating into a series of moans or congratulations.

These meetings can be an exceedingly good means of obtaining feedback. On the whole they are not as well attended as the initial meetings. This is possibly because if IMPACT is going well and the parents are happy, they see no need to come in and talk about it. However, much useful information does come out of such a type of structured and formal contact, and we regard such meetings as an integral part of the IMPACT process.

Summary

This chapter has been concerned with:
- Planning the maths curriculum with IMPACT as an integral part.
- Focusing on the links between 'home maths' and 'school maths'.
- Generating useful follow-up activities.
- IMPACT assessment and record-keeping processes.
- Parental involvement into both of these.

Sharing Maths Cultures

- Designing and sending a good comment sheet or diary for parents to fill out on a weekly basis.
- Addressing some of the issues raised.
- Arranging good patterns of parental contact, including informal chat times and formal meetings.

Chapter 5

Designing and Selecting Impact Materials

We should try to use the children's games to channel their pleasures and desires towards the activities in which they will have to engage when they are adult.

Plato, The Laws

This chapter describes the stuff and substance of the actual IMPACT activities. We describe detailed examples of the way IMPACT works in practice and we give a series of case studies from teachers in Oxfordshire, Redbridge and Barnet. The chapter divides as follows:

1. Designing IMPACT activities.
    a) The criteria for design.
    b) Maintaining a variety.
    c) Preparation in class.
    d) Follow-up work in groups.
2. Selecting IMPACT activities.
    a) The IMPACT resources.
    b) Organization of materials in the staffroom.
    c) Varying the diet.
    d) Preparation in class.
    e) Links with commercial schemes.
    f) Follow-up work in groups.
3. IMPACT flexibility.

We turn to the details of the activities themselves and demonstrate the processes described more generally in previous chapters.

The IMPACT process starts when the teacher is planning his or her curriculum for the term or half term. He or she decides upon his or her topic, consults the Programmes of Study and Attainment Targets, and then considers what resources and materials he or she is going to use. Teachers have a choice when deciding on the appropriate IMPACT activities to fit in with their planned schemes of work. They can design their own shared maths activities, or they can select from the bank of previously developed IMPACT materials.

Sharing Maths Cultures

1 Designing IMPACT Activities

It can appear to be a daunting task to design all your own activities. However, in fact, many teachers see that this can be the easiest way of making IMPACT work for you. Creating your own activities means that each one can serve exactly the role that is most helpful in that week's work. It also gives maximum flexibility and adaptability to circumstances as they change from week to week.

Perhaps the most important advantage to personally designed activities is the sense of ownership which this brings. Materials and tasks produced by someone else, whether we are thinking of a commercially produced scheme or other teacher-drafted resources such as IMPACT sheets, are automatically things which are 'brought in from outside'. The tasks, sheets and activities which we dream up ourselves have a feeling of 'insider-ness'.

To sum up, designing one's own IMPACT activities and writing out one's own sheets preserves professional control and allows the ownership of IMPACT to remain firmly where it properly belongs — with the teacher and the parents and children in the class.

a) Criteria for Design

When approaching the design of three or four weeks' IMPACT activities, it is intimidating if we consider ourselves as sitting with a blank sheet. Ideas seem to vanish and nothing helpful suggests itself. In fact, of course, we do not start with a blank sheet. We have a great deal of information at our disposal already.
- We know our topic.
- We know which attainment targets we intend to cover within the next few weeks, and which aspects of the Programmes of Study.
- We know what maths scheme or worksheets, if any, we shall be using. Hopefully we also have a fair idea as to which parts we shall be focusing upon as well.

How is this information best utilized in deciding what the substance of the IMPACT activities shall be?

1. We need to look at our topic and decide which parts of it could benefit from information brought from outside the confines of the classroom.
2. We need to study the bread-and-butter maths which we intend to do and decide what are its applications in the home, the street or the work place.
3. We need to decide where, if at all, the children might practise or use the skills we are trying to help them acquire in the course of their everyday life.
4. We can think of any aspect of the maths we are trying to do which would be a great deal easier if done at home with the possibility of more individual attention than we are normally able to provide.

We are then in a position to decide which types of activity to send.

b) Maintaining a Variety

It is important to send a variety of types of activity for a number of reasons:
- We want to choose sufficient activities to fulfil each of the purposes outlined above (1–4).
- Some children like one type of activity, some prefer another. Some children love collecting data, others like doing or making something, and yet others really enjoy playing games or working out puzzles. We have to try, not to please all of the people all of the time, but at least to please some of the people some of the time.
- Some activities generate a lot of follow-up classroom activity. Others take a great deal of preparation. We need to balance what we send home in order to make the classwork easier to organize.
- It is important to convey to parents a sense of the breadth and variety of the contemporary maths curriculum. Maths is not just 'sums', nor is it simply games. We have to show by the different types of activities we send what a wide subject maths is.
- We want to ensure that the activities sent involve children in applying a variety of different skills.

Where to start. Activity 1 might well consist of a data collection which starts the topic work for that term. This type of activity makes a good starting point because it:
- involves very little preparation beforehand and so is suitable for the beginning of term,
- can be used to introduce the term's project or topic to the parents,
- generates a great deal of classwork to get the topic going in class,
- depending upon the type and nature of the data involved, can use a variety of skills and therefore can cover a number of attainment targets.

Example. Rubbish collection!

A particular teacher was doing the topic of 'Rubbish' with her second year junior class (year 4). This topic planned to involve the children in analysing different types of rubbish, quantifying amounts of rubbish, looking at policies concerning rubbish disposal, assessing the attempts to 'Tidy Britain', and considering the associated 'Green' issues.

The teacher decided that the best place to start this topic was in the home, with an analysis of the children's own household rubbish. She decided to design a sheet asking the children to analyse what was thrown away in their house on one day at the weekend.

Design issues. When thinking about this the teacher realized that if she arranged the sheet in pre-designed categories, she would automatically preclude the creative thinking of the children in generating their own categories for sorting and classifying rubbish. On the other hand, if she did not supply any categories, it might make sharing all the information back in class, recording and analysing the data, that much more difficult.

If the categories were pre-agreed then it would also be possible to design and arrange the information collecting format for the children. However, the format which is used to generate the data itself affects the data obtained, and the teacher was keen that the children should, if possible, come to realize this for themselves.

Two approaches to this sheet were possible; a very closed and structured approach, or an open-ended and unstructured approach.

This teacher decided that she would compromise and allow the children to generate their own classifications and categories, but provide them with the means of recording the information. This would enable the children to work together and pool the information gained at home, and thereby to agree suitable classifications for household rubbish. The common format would allow for all the data to be shared and processed jointly. In the end she wrote out the sheet as in Figure 5.2.

This was an entirely individual approach to the beginning of her topic work for the term and the IMPACT activity fitted her needs at that time. *How to continue.* The children had such fun with their rubbish weighing and classifying that the teacher mentioned above was drowning in data. This is a common state for teachers after the first week of IMPACT in a term. It does mean that in practice it makes sense for the second IMPACT activity to be one which requires a little preparation but not a lot of following up. It also makes sense, for the balance of activities, that the next activity is one which requires the children to practise a skill or play a game.

Example. Toilet paper tables

A teacher of third-year juniors (year 5) was doing some work as a part of her bread-and-butter maths on tables. She adapts an idea from a book to encourage the children to engage in some tables practice at home. The children are given a sheet of toilet paper (the hard sort!) and a five by five grid. They have to fill out the grid with the numbers from 1 to 10 inclusive, placed at random on the grid. They then have to ask their Mum for a number and pin the toilet paper over their grid. They multiply 'Mum's number' by each number on the grid and write the answer above it onto the toilet paper. They can repeat this with other numbers and new sheets of toilet paper if they like.

Back in class, the children swap sheets of toilet paper keeping their grids. They then have to mark each other's work. This is difficult because they are not given either the original grid or the number by which all the grid numbers were multiplied.

It's all in a name. Many of the children were very enthusiastic about this activity and brought as many as six sheets of toilet paper back into class. This impressed us because there seemed to be nothing particularly innovative about the activity in mathematical or educational terms. We suspect that the activity proved popular for quite other reasons which had more to do with the use of the words 'toilet paper' in the title than anything else!

Where from here? Once the term's work is under way and we are well launched into our topic, what is the next stage? At this point there is a

Designing and Selecting Impact Materials

Rubbish Collection!

Can you sort the rubbish used by your household in one day!

Put a tick in each row every time someone throws away something under that heading. At the end of the day, weigh the amount in each category.

<u>Tins</u>
<u>Old food</u>
<u>Paper</u>
<u>Metal</u>
<u>Cotton wool</u>
<u>Cloth</u>
<u>Plastic</u>
<u>Leather</u>
<u>Wood</u>
<u>Bones</u>
<u>Others</u>

To do this you can either sort the rubbish yourself at the end of the day (YUK!) or you can ask your family to sort it as they throw it!

Weights:
Tins _____ Paper _____ metal___
Old food ____ Plastic _____
Cotton wool ____ Cloth ___
Leather ____ Wood ___ Bones___

Sharing Maths Cultures

Rubbish Collection!

Can you sort and analyse all the rubbish your household produces in ONE full day?

You COULD go through all the rubbish at the end of the day. (YUK!)

OR you could ask people to sort it as they throw it away — the food in one bag, tins in another, paper in another, and so on.....

Write down the categories of rubbish you have used.

Write down which had the most, which had the least and so on....

Can you weigh the amount in each category?

HAVE FUN!!

Rubbish Collection

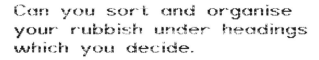

Can you sort and organise your rubbish under headings which you decide.

Look through the rubbish at the end of a day. Decide, roughly speaking, what sorts of different rubbish you can see.
Label each row — Eg. Tins...

Now ask everyone in your family to throw different sorts of rubbish into different bags.

At the end of the day, weigh each bag.

Tins _____
_____
_____
_____
_____
_____
_____
_____
_____

Sharing Maths Cultures

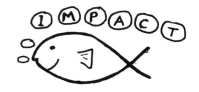

Toilet Paper Tables!

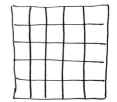

1. Write the numbers from 1 to 10 on to your grid. Each number will fit twice and three will fit three times. Write them anywhere you like on the grid.

2. Pin your sheet of hard toilet paper over the top of your grid.

3. Ask your Mum or someone else to say a number between one and ten.

4. Multiply every number on your grid by the number they give you and write the answer on the toilet paper.

5. Unpin your grid.

6. Repeat this using another number if you like.

7. Bring all your pieces of toilet paper into school!

choice. We can either follow up some aspect of the bread-and-butter maths or we can pursue a different aspect of the topic. Which we do will depend upon the direction we want the classwork to take.

In the case of the topic on rubbish, there are obviously many aspects of the work which cry out for an input from the home. For instance, we can ask the children to look at the wastage involved in packaging. This may involve a calculation as well as measuring and collecting information, so it could link the bread and butter maths work on tables with the topic.

Example. Package wastage
The children are asked to calculate the amount of wasted packing on a product of their choice. They are to look out things which have a lot of extra and fancy packaging, such as toiletries, games or toys. If possible, they are to bring such a packet into school. If they can't then they are to draw it. In either event, they are asked to write down both the measurements of the package and the measurements of the product itself.

The calculation of weights and volumes will be done back in the classroom when the children can examine each others' packets and discuss the dimensions of each one. Some of the volume work will involve the conventional length × breadth × height type of formulae, and will use multiplication. Some will be more practically based and will involve estimation and a certain amount of trial and error.

c) *Preparation in Class*

Designing the home activity can be the starting point for a substantial stretch of classroom maths. Once we have decided what maths the children are going to do at home, we can then think about the preparation which we can do for this in class. The purpose of this preparatory work is threefold:
- It enables the teacher to get the most out of the IMPACT activity before it goes home.
- It ensures that the children know what it is that they have to do at home. This is crucial since if they are unable to explain what has to be done, the burden of the communication will fall upon the sheet itself. In this case, the child is unlikely to do most of the talking. The parent is inclined to do the organizing of the activity rather than vice-versa. This negates the peer-tutoring aspect of IMPACT.
- It gives the IMPACT activity a status and an importance in terms of the on-going classwork which it might otherwise lack. If the maths shared at home is not really a part of the maths in class, then the children are unlikely to remain in much doubt as to which is the most important, in the teacher's eyes at least.

With the activities that teachers design themselves, the tasks and activities preparing the children are very much within the professional control of the teacher. Since the activity has been designed with that particular stage of the

Sharing Maths Cultures

Waste of Packet!

Can you work out how much packaging is wasted?

1. Find something which really has a wasteful package.
Eg. Soap, perfume, games ot toys....

2. Draw the product itself, and the packet (separately).

3. Measure them both.

4. Write down ALL their measurements on your drawing.

5. If you can, bring the package into school.

curriculum delivery in mind, the preparation ceases to be something separate from the general classwork. It would be difficult, if not impossible, even in theory, to specify which pieces of work were preparation for IMPACT and which pieces were ordinary classwork. The IMPACT activity simply becomes a focus for the maths curriculum.

d) Follow-up Work in Groups

The minimum requirement on IMPACT where follow-up work is concerned is that the children are encouraged to share whatever experiences they have had with IMPACT at home, and that they have at least one session working in groups. The follow-up work is essential for the success of any serious attempt to share maths activities at home. It serves the following functions:

1. The children will know that what they do at home is valued and taken seriously if, when they bring the results back into class, these make a difference to the work in progress. There is nothing more discouraging to a child who has brought something from home to be told, 'Very nice, dear. Now you can find your maths book and do some work'.
2. Encouraging children to follow-up what they did at home reinforces the whole process of transferring skills from one context to another. The child takes a mathematical skill which they have acquired (hopefully!) at school. They talk about it and practise it at home. Then they bring the results back into the classroom and take it further.
3. The activities in the classroom which build on the work done at home enable those children who did not get a chance to do their IMPACT at home to share in the experiences of others. All work brought back into class is shared. This included not only data which is relatively easy to share, but also games, puzzles and 'making' activities.
4. The work done in the wake of a particular IMPACT activity is an important part of the feedback from school to parents. Often, parents do not see the point of an activity. The mathematical or educational purpose of something the children have brought home is not always apparent even when parents have listened to their child describing what has to be done, and then have made a genuine attempt to participate. It is important to enable the parents to pop into class and see for themselves exactly where an activity has led and what maths has come out of it.

It is essential that the IMPACT work is followed up in the class and that the children discuss their IMPACT together. If, for whatever reason, the IMPACT activity follow-up work cannot be fitted into a particular week — perhaps the teacher is ill or the children are on an outing or are performing a play — then it may be necessary to give the class a week without IMPACT and a chance to follow up the IMPACT during the next week. Certainly the children and the parents should be clear that what they do at home has an effect upon the classwork and does indeed make a difference.

Summarizing so far. When we are designing IMPACT activities we need to be sure that:
1 They fit with the work we have planned for the children at that time. IMPACT activities should be a help not a hindrance to the teacher. They should reinforce and inform the classwork in ways which are supportive to both the teacher and the children.
2 The children are provided with a variety of mathematical experiences at home. The IMPACT activity should sometimes require children to practise a skill, to collect data, or to make something to bring into class.
3 The actual IMPACT activity is only one part of a process of preparation, shared activity, and follow-up work. It does not have to stand on its own. Therefore it cannot be planned and designed in isolation from the other aspects of the classwork.

2 Selecting IMPACT Activities

IMPACT has been operating now for five years in an increasing number of schools throughout England and Wales. The teachers in many of these schools have been drafting IMPACT activity sheets for most of that time. On IMPACT we collect the sheets and then trial them in other areas and other schools. In this way we have gradually built up a bank of some 3000 IMPACT activities which we know work well.

a) IMPACT Resources

We have structured the materials available to IMPACT into three main categories:
1 those available only to IMPACT LEAs — that is to authorities who have contributed to and are members of the IMPACT National Network;
2 those available to schools in the IMPACT authorities, or to schools who have joined the IMPACT National Network directly;
3 those available to any school or teacher and purchasable from the IMPACT Head Office (address in Appendix).

The materials in each category differ from each other in that they serve different functions. They accordingly vary in both organization and quantity.
1 The LEA Pack. This contains all the 3000 sheets currently available to us on IMPACT. The Pack is updated every term when each authority is sent all the materials which have been designed in each of the other IMPACT Network LEAs. This is, therefore, a constantly increasing and up-dated resource.

The Pack is organized in two main lists. One is a list of all the activities in alphabetical order of titles. The sheets themselves are filed in alphabetically labelled folders. The alphabetical list contains the title of each activity, plus a note of the Attainment Targets and levels to which each one pertains. The second is a list of activities under each level of each Attainment Target from level 1 to level 5. These Attainment Target lists are cross-referenced. For example, a teacher or adviser can look under Attainment Target 2, level 1 and will find a list of thirty or so activities. They can then scan this list to select an activity which also covers Attainment Target 5 at level 1.

This is a very large resource. Authorities which have this Pack have always had some in-service training from IMPACT as well.

2 *The School Resource Packs.* There are two of these and they each contain 200 IMPACT activities. The first one focuses upon levels 1 to 3, and the second one upon levels 3 to 5/6. Each School Resource Pack is organized in the same way as the LEA Pack, therefore each contains an alphabetical list of activities, complete with Attainment Target calibrations. It also contains the Attainment Target Lists of activities under each level, all cross-referenced.

Both the above Packs are only available to the LEAs and schools who are members of the IMPACT National Network.

Information about the IMPACT National Network is available from the IMPACT Head Office (address in the Appendix).

3 *Other Packs.* There are a number of Packs which can be obtained by any school or teacher. These are organized under the following headings:

- Early Years Pack — This is a Pack of some 120 IMPACT activities which have all been designed by infant or nursery teachers. These materials are specifically aimed at the 3–7 age range. They are organized in the same way as the above Packs but they are also listed under a number of general 'topic' headings. This is to enable teachers who work largely in a cross-curricula way or have an easier access to the activities.
- Special Needs — This is a Pack of some 60 activities which are of particular use with children with special educational needs. A few of these are especially designed for children with severe physical disabilities. This Pack is also organized in the same way as those described above.
- Multi-cultural Pack — This contains materials of two types: those which use ideas drawn from non-Western European mathematical traditions, and those which have a general multi-ethnic focus, for example, using other languages or numerals than English. There are 60 activities in this Pack, organized in the same way as those described above.
- Topic Packs — We have produced a series of general topic Packs of materials to enable teachers to select IMPACT activities to fit in with their topic plans. These contain between 30 and 60 activities, and they are listed alphabetically, with their Attainment Target and level calibrations attached. These are available in such topics as 'Travel', 'Ourselves', 'Communication', 'Food and Drink', 'Structures', 'Space', 'Patterns' and so on. This list is not exhaustive.

Further information about the IMPACT Materials, or about becoming a member of the IMPACT National Network, is available from the IMPACT Head Office. The address is given in the Appendix.

Sharing Maths Cultures

ATTAINMENT TARGET 12, LEVEL 1

Animal Data	
Bean Stalk	2i, 8ii, 13i
Coats	
Coin-sort (bubble)	12ii
Coin-sort (pig)	2i, 3i/ii, 8ii, 12ii
Colour of Eyes	
Colour the Squares	
Creature Count	R 2i
Decision Time	2i, 12ii, 14i/ii
Draw a Building	8i, 10i
Favourite Drinks A	12ii, 13i/ii
Favourite Drinks B	R 12ii, 13i/ii
Find 6 Leaves	2i, 8i, 10i
Float or Sink	
Floating & Sinking	R 8i, 13i
Heaviest Apple	2i
Look at your Clothes	R 2i, 13i
Money Coin Collection	2i, 13i
Odd One Out	I 12ii
Rub a Dub	2i, 8i
Shape Collect	R 2i
Shoe Comparison	8i
Sky Fly	R 2i
Sorting Clowns A	12ii, 13i/ii
Sorting into Sets	R 2i
Texture Race	2i, 13i
Things that Move	2i
Three Bears A	2i, 8i
Three Bears B	2i, 8i
Tiny Boxes	2ii, 4i/ii, 12ii
Touch	13i
Washing Day Sort	2i, 13i
What can you feel?	R 2i, 13i

Designing and Selecting Impact Materials

b) Organization of Materials in the Staffroom

The first thing a school — or indeed a teacher — needs to do, having acquired a set of IMPACT materials, is to make these their own. This is, as the Cockcroft Report would put it, 'easier to state than to achieve'. Every school has its own method of doing this, but we have noticed that those schools who seem to have the most thorough knowledge of their 'pack', whatever it is, are those which followed a certain basic pattern:

- A teacher, or more than one teacher, sorts carefully through all the sheets, spreading them out and really trying to classify them for themselves. This may involve some of the headings already given, such as particular Statements of Attainment or Attainment Targets. But it may equally well involve teacher-drafted headings such as 'Early Number', 'Length', 'Area', 'Shape', 'Computation' and so on. It will almost certainly be necessary to photocopy any which belong in more than one category. If the school is using the existing IMPACT headings and lists, then this will mean a great deal of photocopying since the majority of the activity sheets fit under at least two Attainment Targets, and span at least one level.
- The sheets are then put into ring binders under the system of classification chosen or worked out by the staff. Thus one ring binder might have a heading 'Attainment Target 2, levels 1–3'. Alternatively, it could have a heading, 'Basic number work'.
- Some schools or teachers put the sheets into individual plastic covers. The IMPACT sheets get a lot of use with teachers thumbing through them and then taking them out to photocopy or reproduce. Plastic covers are efficient protectors although they can be pricey and therefore something of an extravagance.

Once a teacher has selected the activity she feels will be helpful, she makes a copy for every child in the class, including of course, those who are unable to do the activity and bring it back. She also makes a copy for:

- her own personal records. This is so that she has her own IMPACT file to take with her if and when she leaves.
- the staffroom (this is why the activities are filed in ring binders to which new sheets can easily be added). This means that the school retains a copy of every sheet sent. If teachers write the data and class of children on the back, this provides a very useful record.
- the class record. This is important since it is to ensure that the next teacher will not send home an activity that the children have already done a year or so earlier. Whilst teachers may be able to justify doing the same piece of work twice, at different levels, none the less parents will not appreciate this.

Sharing Maths Cultures

c) *Varying the Diet*

The same arguments which we outlined above under the section on teachers designing their own sheets, apply when they are selecting materials from a pack. The teacher has to look at her/his plans and decide what she/he wants an activity in a particular week to achieve. She/he is then in a position to consult the IMPACT lists under the Attainment Target headings. This process of thinking about the work she/he wants to cover, consulting IMPACT lists, looking again at her/his objectives, studying a number of actual IMPACT sheets, and then maybe re-thinking the planned work, is not a linear process. It can be more easily conceived as a spiral than as a straight line.

Thinking about the maths objectives tells us where to pitch into the lists of Attainment Targets, and studying the IMPACT sheets available generates ideas as to possible preparatory and follow-up classwork. In this way, the choice of materials is not made in advance by someone totally unconnected with that classroom, as it is with a commercial scheme. The National Curriculum structure itself helps us to take care of any problems of continuity and progression. The IMPACT materials help Attainment Targets 1 and 9 to be in the forefront of all the maths covered.

The children also need to practise different skills and to cover different aspects of the maths curriculum. One week they may be doing some arithmetic, adding and subtracting, and the next week they may be measuring. The activities sent during the period of half a term need to include situations

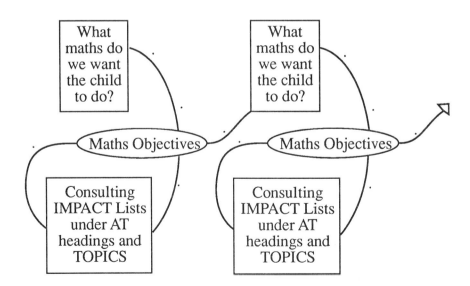

Designing and Selecting Impact Materials

where the children are bringing something into the classroom from the home, and situations where they are practising a skill acquired at home.

d) Preparation in Class

Just as the materials do not themselves dictate which activity is to be sent in which week, so they do not prescribe the work which is appropriate for the preparation and follow-up in the classroom. IMPACT has at times been under some pressure to produce 'sets' of materials which provide some ideas for the preparation and the follow-up to particular activities in the home. However, we have consistently taken the line that the selection, oganization and adaptation of the activities should rest solely with the teacher, rather than being removed from the arena of her/his professional control and pre-decided by those supposedly more knowledgeable or capable.

We feel that the disadvantages of detaching those sorts of decisions from the person at the chalk face far outweigh the possible advantages in terms of coherence and consistency. Only the teacher knows the precise details of the classroom situation in which she/he is operating. Only the teacher has access to day-to-day information about the children's performances. The penalty of arranging activity cycles in ignorance of these details is that the work lacks relevance and at worst can be entirely inappropriate to the precise situation into which it is required to fit.

Example

An IMPACT activity which has been a 'standard' for four or five years is 'How much is your hand worth?'

This activity is traditionally prepared through a selection of the following types of task:
- The children do coin sorts, to make sure that they do in fact recognize the coins.
- The children add up small amounts of money. Some will add only one pences, involving them predominantly in counting. Others will add other coins, and will be doing some more complicated addition sums.
- The children record the addition of coins on paper, either by writing or by colouring in coins depicted on a page.
- The children practise drawing round various things, jam jars, badges, even coins.
- The children draw round various coins and label them. They need to discuss the different shapes and sizes of the various coins.
- The children discuss the notion of 'covering'. They may try to cover a book with matchboxes, or a page with multi-link blocks.

However, a particular teacher might be using the activity 'How much is your hand worth?' in a very different context in which most of these preparatory activities become inappropriate. Suppose that she/he has a class of lower junior children who have had a history of working in workbooks and follow-

070H67
HOW MUCH IS YOU HAND WORTH (B)

Draw around your hand.

Use a 1p coin and draw around it so that you cover your hand.

Count up your coins.

How much is your hand worth?

Try not to leave big gaps between the coins.

If you have got time try some of these:-
Use a different coin.

Use a grown ups hand.

Is your hand worth more than 50p?

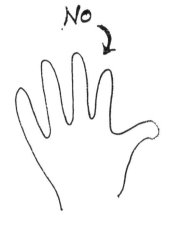

Designing and Selecting Impact Materials

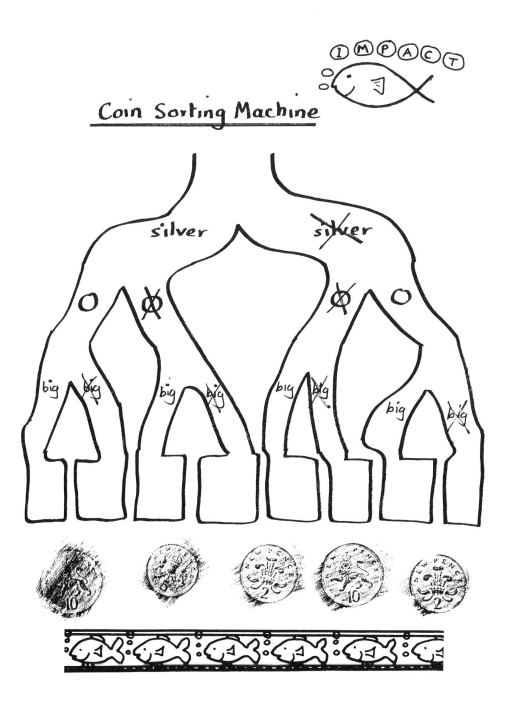

ing a maths scheme in a fairly rigid fashion. The children may be fairly competent at certain written algorithms but the teacher may be aware that they lack other, more practical, skills. For example, they may be extremely poor at measuring and estimating.

This teacher has chosen this activity because it will enable the children to use and apply some of their mathematical skills such as adding and drawing round. She/he also sees that it will require them to make sensible estimates, and to check these in a non-threatening situation at home.

The appropriate preparation in this case looks very different from the tasks outlined above.

- Some of the children lay coins in lines equal to the width of the table, or the length of a book. They have to estimate how many coins there are and how much they will add up to.
- Other children calculate the difference between a book length line of one coin as opposed to another.

The preparation has to depend upon the actual classroom situation and the children themselves, as much as upon the nature and type of the IMPACT activity.

e) Follow-up Work in Groups

The follow-up work will similarly depend upon the factors peculiar to each situation as much as upon the IMPACT activity itself. The teacher may find that a particular activity really takes off and the children and parents generate a great deal of enthusiasm. In this case, she/he will want to exploit the activity and get a lot of classwork out of it. On other occasions, an activity may go down like a lead balloon. For whatever reasons, the shared maths never really takes off and the teacher feels that minimum follow-up of a discussion about what happened is all that is required.

Popular activities. One of the things we have been concerned to monitor over the last five years is the popularity or otherwise of the various IMPACT activities. It has proved much more difficult to find any sort of patterns than we imagined. It is not possible to divide the IMPACT activities into those which are guaranteed to go well and those which aren't. In some schools, or classes, games go very well. In others, games of any kind seem to be about as popular as the back of a bus on a wet Sunday. Even an IMPACT activity which has gone down well in a whole range of schools and classes can suddenly go really badly in a particular situation.

There are too many factors involved in whether an activity goes well for us to be able to predict but such factors include:
- the number of children away;
- the time of term;
- the weather;
- what is going on in the rest of the school;
- classroom factors, such as assemblies, plays, etc.;

- the health and mood of the teacher.

What we can say with certainty is that if something goes well at home it is possible to get the children to do some really excellent maths in the classroom on the back of it.

Example. Tea and coffee survey
The children in one second year junior class had carried out a survey to find out how many cups of tea and coffee their friends and relations all drank in the course of one day.

Some of the children had had great fun with this and had surveyed everyone for miles around. One parent came in and complained that her child had questioned all her uncles and aunts. When the teacher said how nice it was that she had been so enthusiastic and done so much work, the parent replied that it would have been a lot nicer if all these relations had not happened to live in Australia!

The teacher was really able to capitalize upon this. The children not only recorded all their data, both on the computer and by making a huge block graph, they also carried out a similar survey of all the staff in the school. They then used the information to calculate how much coffee and tea various people drank in a week, or a fortnight. They worked out that the staff of the school drank a whole bath-full of coffee between Monday morning and Friday evening! We felt that this was not really something that the staff needed to know!

When an activity goes really well, the subsequent work in class is also likely to go well. A wise teacher will adapt her/his plans to take account of this, since otherwise good opportunities will be lost.

f) Links with Commerical Schemes

The IMPACT activities fit with any scheme or set of books. The teacher can plan the children's work within the scheme in conjunction with the preparation and follow-up for IMPACT. However, the work that children do on a worksheet or in an exercise book while working through a commercial scheme cannot replace the work which forms any part of the IMPACT activity cycle. This is because there are important differences between the sort of mathematical tasks which children do as a part of a scheme, and those which form the preparation or follow-up to a shared activity in the home.
- The preparation or follow-up work always involves talk. The children are encouraged to discuss what they are about to do or have just done at home.
- The preparation or follow-up work is only occasionally individualized. Either a group of children or the whole class will take the same activity.
- Where possible, the activity at home and the follow-up work in the classroom will be responsive to the children's interests. Since IMPACT

Sharing Maths Cultures

Tea/Coffee Survey

How many cups of tea and coffee do you and all your friends and relations drink each day?

Help us in our survey by writing the names of as many people as possible on the chart below.

Name	Number of cups of tea in day	Number of cups of coffee

Who drinks the most?...............
Who drinks the least? ....

activity cycles are planned rather than determined, they retain a flexibility which a scheme cannot.

One of the great advantages as far as we are concerned about keeping the IMPACT Pack separate from commercially produced materials was that the IMPACT materials will continue to fit with whatever scheme the teacher is using. The IMPACT process is unaffected by how the bread-and-butter maths is organized.

3 IMPACT Flexibility

In the last three chapters we have discussed the IMPACT process stage by stage. The basic unit of the process is the activity cycle. Although we are at pains to outline the process itself, we are equally concerned to point out that there is no step-by-step procedure to any part of this process. We do not prescribe *how* the activities are to be prepared, chosen or designed, or followed up in class. We do not specify the means by which a comment from the parents and the children should be obtained, nor the methods for arranging meetings.

We do, however, draw the parameters. We are clear that the IMPACT process commits teachers to doing certain things, but it does not tell them *how* these things must be done.

Advantages

a) *Different levels, different activities.* We have spoken throughout as if it were always the case that a teacher will select one activity for the whole class. In practice of course, this will not always be the case. The advantage of a flexible approach is that the basic process remains the same, with the idea of the activity cycle at its core, even when it proves necessary to adapt the details.

The number of activities which a teacher needs to send to her class will depend upon two things:
- the type of activity,
- the range of level within the class.

i) Type — Data collecting

With a data collection activity, where the children are finding something out, and are bringing some information back into school, the same activity will almost always do for all the children, whatever level they are operating at. It is a good idea to write a couple of lines onto the bottom of particular children's sheets if you feel that they are being asked to do something a bit easy.

For example, if the children are being asked to perform a traffic survey, some of them can record it in a more systematic way than others.

Traffic Survey...

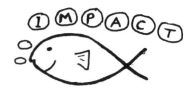

Can you ask a grown-up to take you to a place where a fair bit of traffic passes.
This could be a room in your home, or it could be a nearby street.
Stand somewhere safe.

Count all the traffic which passes by putting a tick in the right row.
Do this for ten minutes.
Ask the grown-up to time you.

Car_____
Bike_____
Van_____
Lorry_____
Ambulance_____
Fire-engine_____
Motor-bike_____

Designing and Selecting Impact Materials

Traffic Survey....

Can you ask an adult to take you to a place where there is a fair amount of traffic. This could be a room in your home, or it could be a nearby street.

Stand somewhere safe.

Count each thing that you can which passes by putting a tick on top of one of the pictures.

Do this for ten minutes. (Ask the adult to time you.)

Can you count how many ticks you have in each line?

Thus the same basic idea will serve for children of different ages and stages of learning.

ii) Type — Doing and making

Once again, this type of activity can usually be the same for the whole class. It may sometimes be necessary to modify some children's sheets to make the task slightly more simple or more complicated.

With both the data collecting and the doing and making, the difference in the level at which the children are operating is taken into account in the work which happens back in the classroom. When the children bring the products of the home activity back into class, different groups of children will handle the data, or work with the information in different ways.

iii) Type — Games, puzzles and investigations

It is with the activities that rely most heavily upon skills practice that it is often necessary to send more than one sheet. Sometimes one basic activity will do, with different adaptations for different levels.

On other occasions it is easier to send different activities, although these will quite often have the same format.

The advantage of using the same format is that it is less immediately apparent to the children that there are different levels of activity.

The best solution to the problem of a class of children who are operating across a wide range of levels is undoubtedly to select or design, as far as possible, those activities which are themselves 'elastic'. That is to say, they can be approached at a number of different levels. There are many activities in the IMPACT Pack which are of this form. Investigations can be particularly helpful here.

b) Non-participating children. A question which is always asked whenever we are talking about IMPACT is that of what happens to the children whose parents never participate. There will not be a large number of these, since it is possible for 80 per cent of the children to be taking part in *any* school. However, the fact that there are only three or four children in a class who never do their IMPACT at home does not reduce the magnitude of the problem with respect to them. Teachers have adopted a number of strategies to enable these children to get the most out of IMPACT.

1   All the data brought in from home is shared as a matter of course. Exactly how this sharing takes place can be so structured as to particularly include the non-participants. For example, a group of children including a couple of children who have not done the IMPACT activity and three or four who have, can work on the data together.

2   Extra adults in the school can be utilized as 'IMPACT' helpers. 'Go and ask Mrs Brown if she wouldn't mind letting you draw round her hand after she's finished clearing away the dinners'. This was said by a teacher to a group of three children, one of whom had done the IMPACT activity at home and so knew what he was doing, and two of whom had not. There are many adults around the school who do not mind answering

Make-a-shape

Can you cut out and fold up the shape below to make a strange-shaped box!

Be careful not to cut off the tabs. You need them for gluing your box together.

Make-a-shape

Can you cut out and fold up the shape below to make a strange-shaped box!

Be careful not to cut off the tabs. You need them for gluing your box together.

Before you fold your box, can you put a number on each face?

Now can you draw the shape you would have to fold to get a cube? Look carefully at a dice to help you....how many numbers does it have?

NUMBER GRIDS

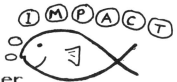

1. Throw your dice and enter the score in the grid.

2. When you have filled in the grid, add the numbers horizontally and vertically.

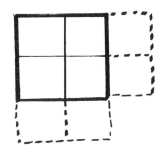

3. Add your answers both horizontally and vertically. What do you notice?

4. Enter this final number in the corner square.

Extensions

Does this work with all numbers; odd, even, 2-digit, 3-digit, etc?

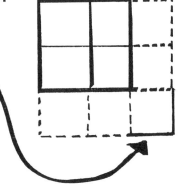

Try different size grids.... 3 x 3 or 4 x 4?

What if the grid is not square?

Use a calculator if you want to try very big numbers.

Sharing Maths Cultures

015D032

Dinosaur Game (A)

Throw the dice in turns.

Match the number of dots on the dice to the number on the dinosaur.

Cover up that number using a counter or button or lego brick.

The first player to cover all the numbers on their dinosaur is the winner.

Make your own game by drawing your own picture and putting your own numbes on it.

Designing and Selecting Impact Materials

015D033

Dinosaur Game (B)

You will need someone to play with, two dice and some counters.

Throw two dice.
Add the two numbers together. Take 2 away from this number.
Cover the right answer on the dinosaur.
The winner is the person who has covered the most numbers.

Designing and Selecting Impact Materials

Dinosaur Game (B)

questions about how many cups of coffee they drink, or having their head circumference measured! It is often good if the children doing the activity with the kind school helper include one or two who have done the activity as well as those that have not.

3   Sometimes, older juniors can be invited to come into the infant or lower junior classrooms and work with some of the children who haven't had a chance to do the IMPACT activity at home. If this is a regular fixture, the older children may form a good relationship with the younger ones, and can come to have quite a proprietorial attitude towards their progress. There is also the opportunity for introducing peer- and cross-age tutoring. (*See* Topping, 1988.)[43]

4   Within the follow-up classwork it is possible to make sure that some children do another version of the IMPACT activity. This can be helpful both for some of the children who have shared the maths at home, as well as for those who have not managed to do so. So, for example, if the children had done a 'How much is your hand worth?' activity at home, a group of children could do a 'How much is your foot worth?' in the classroom. This enables those children who have done the activity to 'tutor' those who have not. This is beneficial for the learning of both sets of children.

It is always important to remember that in any one week there will be two sorts of children who have not done their IMPACT. There will be those who have not done it that week, but who do share the activity at home sometimes. Then there will be those, a much smaller group, who never share the activity at home. This means that the effort to give the non-participants a shot at the activity in school-time does not have to be always targeted at exactly the same group. There may be constant members, but many children will float in and out of it.

c) *National Curriculum or not.* The National Curriculum has, almost overnight, altered the way in which teachers are approaching aspects of their work. IMPACT started in 1985, before the National Curriculum was even a twinkle in Mr Baker's eye. However, despite the fact that it has continued to operate through a period of rapid educational change, IMPACT has proved sufficiently malleable to survive and even to expand. If IMPACT had prescribed certain rigid procedures, then these would have been rendered useless or inappropriate by the changing circumstances. By relying upon a philosophy and a simple process, teachers and schools have been able to adapt their own methods to their own changing circumstances.

Summary

IMPACT involves a very simple way of life.
- Inset workers or other teachers *plan with* teachers.
- Teachers *work with* the children.
- Children *work with* the parents.

- Parents *evaluate* what is going on.
- Teachers *take account* of the parents' evaluations in their records.

The materials, i.e. the sheets on which the IMPACT activities are outlined, are an intrinsic part of this process. They provide the vehicle for the activity in the home. It is crucial that the IMPACT sheet for each week is selected or designed by the classroom teacher with all her/his own specific needs and criteria in mind. The resources themselves are the end point and not the starting point of IMPACT.

Further information about IMPACT Packs of materials is given in the Appendix.

Chapter 6

Evaluation: Reflecting on Practice

'On a waiter's bill pad,' said Slartibartfast, 'reality and unreality collide on such a fundamental level that each becomes the other and anything is possible, within certain parameters.'
　　　　　　　　　　Douglas Adams, Hitch Hikers Guide to the Galaxy

Re-thinking the Scope of Evaluation

In a, perhaps apocryphal, film version from the *Flash Gordon* canon, the Saturday matinée serial of the 1940s, the heroes, Flash (the brawn), Dale (the beauty) and Professor Zarkoff (the quirky man of science) land their spacecraft on the planet Mongo. Having disembarked, stretched his legs and taken in the scenery, the Rasputin-like professor imbibes a long deep breath and passes the following inductive comment, 'Hm good,' he says, 'There's oxygen on this planet.'

Well, that'll come in handy, one thinks — before going on to think, 'What an accurate nose science gives you'. Meanwhile, Flash and Dale have found other things to be getting on with, hardly having stopped to eavesdrop on Zarkoff's asides to the audience.

Once you are already in the middle of doing things it seems unnecessary to stop and produce irrelevant comments that appear to come from outside the situation in which you find yourself. When we stop to reflect on what we are doing we seem only to produce statements which, to some extent, we 'already know to be true'. We sometimes find ourselves checking or testing, say, a child in order to produce comments that are designed to persuade some 'audience' to suspend disbelief in our practices rather than actually change what we are doing. Like having oxygen around, we are far too dependent on some of our habits either to inspect them too closely or discontinue them. After all, as teachers, we *know*, don't we, how things are going. A fear of criticism, especially when directed at ourselves, has led to an evasive use of evaluation tools. Evaluation, if allowed to focus on us, will, we tend to think, reveal unpleasant gaps in how we teach. Rather than that we prefer to keep evaluation at arm's length. Strangely enough, this leads us to permit its use of pseudo-scientific terminology, to let the boffins and the professor Zarkoffs of education have their weird vocabulary to describe what we do. And we allow all this to reduce the potency of evaluation and to make it something outsiders need rather than us as insiders. If they produce figures that do not fit in with what we know to be happening then we tend to think there must be something wrong with the way they arrived at them.

From Guilt to Reflection

That evaluation is such a long way off from what we take to be real concerns, a bureaucratic add-on of endless form-filling that sits on other people's desks, in itself is a sad, if revealing, comment on how we conduct our affairs. Part of the problem seems to be that we incline to guilt or blame more easily than looking reflectively at ourselves. We ought to consider whether we use evaluation to confirm our suspicions, to locate blame, or to engage in penitential self-deprecation.

The problem appears to have to do with what we look for as a matter of course when employing evaluation tools and what we perceive to be the main factors which we want to investigate. Schools are complex social entities, yet we appear to find it relatively easy to locate problems in particular children, in parents, in colleagues and even ourselves. We do not claim that this is all delusion, far from it. But we think it is not the whole story. There are other things we ought to take into consideration when producing reflections on practice.

In the introduction we quoted Wood (1986) discussing teacher–learner interactions. He says that these interactions are 'tightly constrained by the nature of the institutions that we have invented to bring teachers and learners together'. This perspective on what happens in classrooms and schools is one we find immensely difficult to see while actually engaged with children in the classroom. We are less likely to 'find' any tight constraints or such like when evaluating what we do than we are to find problems with individuals: children who have not progressed, teachers who adhere to schemes, heads who spend money on the wrong things.

Inviting parents to read at home with their children as part of school practice now seems obvious to us. But it had been a momentous *institutional* change when schools up and down the country started doing it. Prior to PACT we had been locked into the narrow focus: blaming ourselves, the children or their home background for deficits in reading ability. The imaginative leap which produced a change at the social level to improve standards at the individual level started with looking beyond our noses: that is, beyond the criticism of others as well as, and perhaps more importantly, the criticism of ourselves as teachers.

Having now set the scene for evaluation slightly wider than usual we now turn to the question of evaluating, relating evaluations to complex situations.

Evaluation in Complexity

Being creatures of adaptable disposition, we often find ourselves changing our minds in the middle of doing something. Then, later on, we become confused about what we feel about what we did. Going to a bookshop with a list of titles to buy, finding none of them and returning home with a pile of thrillers which we end up enjoying immensely can give us ambiguous feelings when evaluating our expedition.

Dinner parties are always a bit of a worry. Presumably you do not pass around evaluation forms for your guests to fill in afterwards; yet you are anxious about *how it went*. Well, maybe the soufflé dropped a little, but the chocolate mousse was something to pat yourself on the back for. Imagine if you *did* pass forms around after. What would you like to ask your guests about your party? You could start with the food — too cold, overcooked, not enough; the conversation — dull or inspiring; the lighting — too bright, etc. If you now think about what you do when you come away from other people's dinner parties you may find any such form inappropriate. And yet, as you get in the car you comment on how much that person that Graham brought with him talked; or how Kate got drunk; what a sparkling conversation you had about Dutch pancakes, etc. All these things could not be predicted in advance (except, possibly, Kate drinking too much) and so could not go down on the form. In many ways, the things we know about in advance may bear only tangentially on the success of the evening. On the other hand a dinner party without a dinner would be grotesque. The dinner can be inconsequential to what one felt about the evening: the next morning you remember what you said but not what you ate. On the other hand if the food had been desperately awful it may be all you, or your stomach, can recall.

For evaluation purposes this should alert us to the complicated relationship between:
a) what is necessary to *establish* a situation — to make it happen at all (food for dinner),
b) what emerge as *effects* from the situation — that make it the special situation it is (discussion on pancakes), and
c) what we *experience* in the situation (a good time).

We attend dinner parties for food and conviviality. But we cannot evaluate either alone. And we cannot evaluate both on the same basis. Schools have more in common with dinner parties than laboratories, but they are not strangers to evaluation.

Knowing What to Evaluate

It is insufficient when doing something like IMPACT in a school to evaluate on the same basis as you would a more focused intervention such as 'cued-spelling'. Cued-spelling stipulates a defined set of procedures for peers to tutor each other in spelling. It is directed at specific pupils at specific times and the skills involved are not so numerous that they cannot be listed and assessed as part of a planned monitoring exercise.

IMPACT will not be quite the same. Different elements of what is, after all, a broad institutional change in a school will have numerous emergent effects. Any of these at different times can become significant depending on the occasion, while others, as with dinner parties, will recede in importance. We suggest that you might proceed along the following lines.

1 Accommodation for IMPACT in school
Examine how IMPACT is to be accommodated into your school. Look at what changes and adaptations need to occur and list them after discussion.

2 The classroom
Teachers might pinpoint difficulty areas in the class or particular children and look at strategies for relating to parents: when and how contact might be made.

3 Decide an IMPACT focus
Decide why IMPACT is to be adopted in the school — what particular *focus* it will have with respect to such things as:
- getting more parents into school,
- changing staff attitudes,
- changing parent attitudes (to maths, to the school),
- changing children's attitudes (to maths, for example),
- bringing in more practical maths,
- introducing cross-curricula maths work,
- improving mathematical attainment,
- improving staff expertise in the curriculum,
- implementing the National Curriculum,
- changing traditional school–home relations,
- other areas or issues on which to focus.

For each of these you will need to decide what success will look like for whatever you define as aims. We shall have more to say about this later.

4 Emergent effects
Like dinner parties there will be unexpected prizes and disappointments. These need to be recorded and discussed at parent and staff review meetings.

5 Monitor and record
You will need to monitor and keep records. We discuss this in detail below and it has been raised elsewhere in this book. You should be aware that monitoring should be part of your establishment of any intervention and woven into the structure of your practice.

6 Focused assessment

Periodically you may wish to make specific assessments to test the waters. The National Curriculum has its own built-in testing arrangements, but there is nothing to stop you following up anything that particularly interests you with regard to such things as mathematical attainment or the amount of time devoted to practical maths.

Rationale for Monitoring

We are surprised to see ourselves on video and are inclined to giggle at, if not disbelieve, what we see. When working having had little sleep we are apt to assume we look a mess — this may be unfounded. We may spend considerable time with our cosmetics and assume we look wonderful — this may be, alas, also unfounded. Over longer periods we are inclined to view the past through rose-tinted spectacles and think we see 'good old days'. We are also subject to mood swings. Making statements about how things are now in comparison to how they have been is spurious without recourse to records of some kind.

Of course monitoring of any kind cannot 'capture everything', but that is not its purpose. It is there to give us a sense of our activities or practices from a perspective outside our attention spans. Halley's comet reappears every 76 years. If records did not monitor heavenly activities of this kind we might assume that each sighting of the comet was a unique event and not the return of the same object.

Education abounds with examples where we can spot regularities in children's behaviour and link the regularity to some other correlated happening. A frequent, and regular, absenteeism could be attributable to the recurrence of a certain lesson with a certain teacher on the time-table. Such things are fairly easy to spot. With IMPACT the reasons why, say, games are correlated with greater parental response in one class and the marked opposite in another need further sustained scrutiny: and that means monitoring and looking deeper.

Monitoring extends the facility of our memories to access events in ways which we might have at the time but did not because then other events seemed more interesting.

Assessment and Controlled Studies

It is a good idea to look at mathematical attainment narrowly, or as part of an assessment of integrated curriculum practice. With available assessments you will need to read the instructions on the tin. But you will also need to remain vigilant about what you are assessing (*see below*). Assessments are confidence boosting, or at least give one a sense of where one is falling short.

But great care needs to be applied so that what we think we want to know is being addressed by the assessment.

Conducting controlled studies to examine the worthiness of an intervention is a waste of time unless you have a specific focus for the study such as 'Do five 20 minute sessions with a skills-practice game improve coin-recognition in six-year-olds?' Arriving at such questions is the important process and can form part of your overall agenda in introducing an intervention like IMPACT. The value of such exercises tends not to reflect on the worthiness, or otherwise, of the intervention as a whole — but can and should be used for re-focusing changes in specific practices.

Techniques for Controlled Studies

The first thing to say is that the techniques used in experimental studies require some practice. People using them for the first time tend to be finding out as much, if not more, about what doing controlled studies is like than finding out about their area of enquiry. This applies also to the statistical techniques which are employed along with the overall design of the study. Many use 'cook-book' statistics and all too often get to the end of an analysis and do not know how to interpret their results back into educational terms. Skills for doing controlled studies are not difficult to acquire and the world of teaching has many persons who are capable of sharing these skills.
Formal observation. We have already discussed, in chapter 2, the nature of the relationships between the mathematical techniques, like adding and subtracting, deployed by children and the social settings in which they occur. Any professional strategy intent on gaining deeper insight into learning situations and the difficulties that children meet there needs to focus on some aspect of that social setting. We set out in chapter 2 a non-exhaustive list of possible foci for attention. These were:
- TECHNIQUES — mathematical operations that produce habitual performances from children when used.
- CHOICE OF APPROACH — instances where children see techniques as 'units', appropriate or inappropriate given their reading of the circumstances.
- AN INDIVIDUAL'S LANGUAGE FOR MATHS — how a child uses language within mathematical work.
- DIALOGUE — a rich source of information about how sense is being made of mathematical projects and how the above are coordinated.
- SOCIAL CONTEXT — the larger settings within which maths happens has a lot to do with the structure of children's mathematical undertakings.

These elements are intimately related, but adequate attention needs to be paid to each. Remember that when we examine human encounters in depth and detail we are not necessarily focusing on what children or people do or how they perform *individually*. We are interested in looking at how, for example, dialogue organizes the way we make mathematical events happen.

Sharing Maths Cultures

We have to get a sense of how a child becomes engaged with another, or with ourselves as teachers. Looking more closely at this engagement we might be able to discern how our approaches and performances create the contexts within which a child operates.

Techniques for looking at the Early Years conversations are given in Wood *et al.* (1980).[44] Methods of looking at classroom 'discourse' are given also in Stubbs *et al.* (1979).[45] Practising looking at familiar situations in unfamiliar ways gives us greater insight into what we are doing and often rejuvenates our teaching.

We now wish to turn to IMPACT giving examples of evaluation issues as they have arisen and what they meant to those involved. We then move on to consider how we can begin to get further insight into these issues; and finally we look at IMPACT evaluation and National Assessment.

Starting with Intuitions

How people feel about IMPACT is a good indication of how it is going. Rather than asking about one arbitrary aspect — 'How many responses are you getting?' or 'Have you managed to stick with the activities you planned?' — we find that starting with how the teacher is feeling produces useful and illuminating information.

Example: a teacher
In a particular classroom, the teacher is feeling depressed about IMPACT. She says that she doesn't feel it is going very well. In further conversation, it transpires that only about 18 of the 27 children are responding, whereas to start with nearly all of them were doing it at least some weeks.

The teacher herself is clearly using response rate as an indication of how well things are going. First we should look to see if there might be other indicators. For example, how are those children who *are* responding getting on? The teacher is much more positive here. She feels that they are getting a lot out of it, and one child in particular really seems to have gained in terms of her confidence and understanding in maths. We then turn our attention to the reasons why some of the parents are no longer responding. We discuss doing an IMPACT assembly to which parents are invited. This might remind parents how useful it can be if they do share the activity.

Parents' Intuitions

Just as the teacher's feelings make a good starting point for evaluative procedures, so the parents' feelings can do the same. But we have to distinguish here between the teacher's perceptions of the parents' feelings and the parents' own expression of these. Sometimes a teacher will be unduly affected by the feelings of one or two vocal and exceptionally persistent parents, and will take their feelings as representative of the whole body of parents.

We have, on occasion, gone into a school and talked to the teacher about how IMPACT is going, and been told that the parents all think that the activities are too easy and don't see the point of it. Upon talking to the parents and reading carefully through all the response sheets we find that it is actually only a relatively small number of parents who feel this way. However, those that do are particularly articulate and tend to be those who write the most.

The starting point where parent's feelings are concerned should be the comment sheets and the responses made at parent meetings. It is important to look carefully at these and to count up the number of responses of each type.

Children's Feelings

Sometimes it seems as if in discussing home–school liaison, we spend a great deal of time thinking about how teachers can listen to parents and take seriously what is said, and how parents can come to see what a teacher is trying to achieve through a particular series of activities or tasks. But we almost forget to ask or hear what the children have to say!

If we go into a school and ask the parents how IMPACT is going, and they seem very enthusiastic, and we then talk to the teachers who seem equally so, it can appear that everything in the garden is lovely. But if the children are unhappy, then several important parts of the IMPACT process are not in fact working as they should be. If the children describe IMPACT as boring or too difficult then it is unlikely to be the case that it is improving either their confidence or their attitude to maths. It may also be unlikely that they are acting as the 'instructor' in the home, and setting up the activity.

When children have complained to us about IMPACT, the substance of what they are saying is usually that their parents don't listen. This is more important with some activities than others, since those such as games, can automatically involve more genuine dialogue than those such as a making activity.

Monitoring IMPACT in the School

We need to establish a system for monitoring that allows us to respond to our findings. The following points are useful.
1 Questioning current effects.
2 Looking at the teacher's role.
3 Where are things going well and badly?
4 Consulting others: colleagues, parents, children.
5 Deciding whether to make changes.

6 How to change.
7 Looking at effects of change.

This process will inevitably involve two things: making judgements and collecting evidence. Of course these are not separate but intimately related. Not only do we make judgements about how things are going, but what counts as evidence and what does not is itself a matter of judgement.

1 Questioning Current Effects

i) in the classroom,
ii) at home — through comment sheets,
through discussions,
through meetings.

i) in the classroom. Each classroom is obviously its own micro-world, and what is a disturbance in one classroom is a matter of normal routine in another. Each teacher has a sense of how her/his own class is doing. Bad days, good days, a whole good spell, and then a not so good patch — there are signs that the teacher recognizes without even being necessarily conscious of the reasons behind them.

Relevant questions might include:
Do the children seem enthusiastic about taking the IMPACT home?
Do you think they are keen to show you what they have done at home?
How many of the children are taking part?
Do you have difficulty thinking about how to follow up certain activities?
Is the shared maths a help, or a bother as far as the classwork goes?
How many of the parents and children comment on the activity?
Has anything noteworthy, i.e. particularly good or bad or unusual happened?
At the same moment that we start to describe, we consider reasons and causes. We wonder, and we try to come up with explanations. It is this process of reasoning which can help us to generate the categories which we will eventually use when it comes to evaluation.

ii) at home. As teachers we do not have direct access to how IMPACT is going at home. We have to rely upon the various means we have set up by which parents and children can make their feelings plain.

a) Through comment sheets

The comment sheets which accompany every activity are an excellent way of obtaining feedback. Most of the information obtained through the comment sheets or diaries concerns how the activity went and is provided in well-defined catogories. This means that we only get the answers to the questions we have thought of asking! What do the parents choose to talk about? Why do they focus on one thing rather than another? To what aspect of the process do their questions relate?

b) Through discussions

Both parents and children will on occasion have some fascinating anecdotes about IMPACT. Sometimes it is something which went really well, and on other occasions an activity was misunderstood or misplaced. These stories tend to emerge in the situations encouraging informal discussion — class teas, short ten-minute chats with the teacher and so on.

Sometimes, it is easier for children to talk about what happened as they were sharing the maths at home when they are discussing their comments with the teacher. On other occasions, it may be that through the follow-up activities, stories about the part that was done at home will be related.

c) Through meetings

The type of feedback obtained through meetings tends to be very much more formalized than that obtained through discussions and even the comment sheets. The groups of parents report back to the whole meeting with three positive points and three things which they believe need improving. The resulting list, usually of about twelve things in all, is kept as a record of that particular meeting. In this way it is possible for a teacher to monitor how the parents' responses change over time. Looking back over the records of the meetings over a three-year period does present a very informative view as to how the parents' preoccupations change.

As IMPACT meetings become more of a routine part of school life, and less of a novelty, the parents who attend regularly start to be quite relaxed about sharing stories of amusing IMPACT events. Often different activities will be discovered to have been treated quite differently in each home.

2 Looking at the Teacher's Role

Having thought carefully, discussed and made notes about the effects we wish to address, we then consider what role we have in each aspect we can identify without our control.

For example, it is up to each teacher how they treat the work the children bring back from home. Whether the children's work is taken with a throw-away remark, 'Yes, very nice, dear. Put it on my chair and I'll look at it later!' or with an enthusiastic eulogy of praise, is a matter which is clearly within the control of the teacher.

By contrast, there are aspects of IMPACT over which a teacher has no control at all. She/he may do her/his best to encourage the parents to comment on what happens at home, and when they do so, she/he can show that she/he values their remarks by taking them seriously. None the less, at the end of the day it is the parent who will decide whether or not they actually put pen to paper and fill out the response sheet.

3 Where Are Things Going Well and Badly?

Example. Confused children, unhappy teacher? A second-year junior teacher was convinced that, although she conscientiously tried to prepare the children adequately for their IMPACT activity each week, none the less, she felt that the children didn't really seem to understand what it was they had to do. They were not a good class when it came to 'listening' and the final session on the rug on a Friday upon which so much depends always seemed to her to be more of a mess than anything. The session should, in theory at least, be where the teacher conveys enthusiastically what the children have to do at the weekend. They should depart with the calm knowledge and a keenness to get started. By contrast, this particular teacher felt, her little darlings had had to be told off for not listening so many times, and had been generally so disruptive that it was difficult to say that anyone remained enthusiastic about anything!

As far as the parents' and children's comments were concerned, the parents mostly did not fill out the comment sheets, and therefore the children tended not to do so either. Most of the children did share the activity at home, and the weekly response rate in terms of maths returned was high, but the comment sheets were not seen as a necessary part of the process.

However, another teacher, who happened to teach a good number of these children's younger brothers and sisters, saw a lot more of the parents. Her class were middle and top infants, and therefore the parents were normally there to collect their children in the class or at the door. This teacher had time to talk to the parents before and after school each day, and she frequently found herself discussing, not just her own class's IMPACT but the IMPACT set by the first teacher as well.

Her reaction to the first teacher's feelings about the children being under-prepared and demotivated was that this did not at all accord with what the parents conveyed to her. She felt that most of the parents were very positive about the way that their children understood what had to be done, and set about organizing them all to get on with it.

In this case the teachers agreed to get together and have a couple of afternoons when they read a joint story and laid out a cup of tea for the parents to try to get some more informal feedback.

A summary of some questions to ask
- How do I, the classroom teacher, feel about it?
- Which parts do I feel best about, and which parts worst?
- Are there bits which really feel like a lot of extra trouble or work?
- Do any aspects give me a real sense of satisfaction and of something worth while?
- Are some parts almost dreary and boring?
- Do some children really seem to gain a lot?
- Are other children really seeming to miss out?
- Do I feel able to compensate to some extent?

The starting point here is the teacher's own subjective impressions. None the less, we need a considered view of these. This is not a case of noting an immediate reaction and building a great deal on it. One swallow doesn't make a summer, and one bad week of comments, or conversely, one excellent week of comments, must not be allowed to have undue influence.

4 Consulting Others: Colleagues, Parents and Children

An important part of moving away from a subjective impression is to obtain another angle on the project by finding out what others think. This is not the same as listening to parent meetings or reading the weekly comments. This involves:
- deciding what parts we think are going well and why,
- deciding which parts are not doing so well and why,
- drafting some questions about those aspects which can be asked of other people who are also involved to find out if these impressions are correct.

There are three sets of people who could appropriately be consulted here:

i) Other teachers. The easiest way for a teacher to discover if her/his own subjective impressions are correct is to discuss these in some detail and in a relatively relaxed fashion with colleagues. This may be over lunch once a term, or it may be at an IMPACT staff meeting or even in the pub(!), but the resulting information is an essential stage in obtaining another perspective.

ii) Parents. Having decided upon what she/he thinks is going well and badly, the teacher can see if the parents' perceptions agree with hers/his. The easiest way to do this may be by means of a small questionnaire, or through informal chats, or consulting the data from monitoring to date. The situation of the school, the age of the children, the numbers of parents who are in or around the classroom each day, are all factors which will help to decide how best to obtain the parents' views.

iii) Children. Teachers, like parents, are sometimes not the best people to talk to children if what they want are their honest opinions! There are a number of reasons why what the children will volunteer in class, in even the most open and frank discussion, may not necessarily be useful. Some children will be tempted to say what they imagine the teacher wants to hear. Other children will say precisely the opposite of this. Of course, this in no way invalidates either the purpose of holding such a discussion, nor the information thus obtained. But it may mean that it is useful to try to get another angle on the children's views.

One method of doing this involves asking a non-teacher friend to come and talk to the children about IMPACT. On the Project, we have frequently served this role ourselves, coming into the classroom or the playground and having a long chat with various groups of children about how they feel IMPACT is going; what bits they particularly like, and which parts they feel unhappy about. What we have learned is not infrequently in conflict with the teachers' own impressions of what the children think.

Sharing Maths Cultures

You may decide that this might be a point where, if applicable, a controlled 'study' or focused assessment might be appropriate — once, that is, you have explored other perspectives and have isolated agreement on all fronts.

5 Deciding Whether to Make Changes

It is always assumed that one of the purposes of an evaluation is to affect policy and practice. This may be true, but it must be remembered that leaving things as they are, or making a positive decision to continue with certain practices, is itself as important a result of the evaluation as the bringing about of change.

It is important to consider very carefully before deciding to make changes to the practices and routines of IMPACT. First of all, it takes a while for the procedures to become routinized and if they are subject to change, this process of routinization and familiarization is considerably slowed down. Secondly, in order to make a change we have to be fairly sure that we have correctly identified the cause of a particular problem and that the envisaged change will assist in solving it.

6 How to Change

Once we have established that it seems a good idea to alter practice in a particular way, we can set up the means of doing so. This will inevitably involve first of all informing everyone concerned that something or other is going to alter.

Example. Changing the day. A teacher had carried out an evaluation of the way IMPACT was running in her class and come up with the following conclusions:

- The children generally felt that they often didn't quite understand what they were to do at home and that they would like a bit more help with it.
- The parents felt that the children were often uncertain and that the instructions sent with the child were too hard to follow, especially given the maths-phobia of some of the parents!
- The teacher felt that she got a lot of very positive responses and that the children, and the parents, were doing some wonderful work at home. She saw that they were somewhat lacking in confidence when it came to interpreting what had to be done and that they were very conscientious on the whole about completing the activity over the weekend.
- Overall, it was felt that IMPACT was helping the children's learning of maths quite considerably, and that some of them were really gaining a lot. The parents were enjoying IMPACT, and felt it to be worth while,

and the effect of the work done at home on the subsequent week's classwork was remarkable and very positive.

In order to address the few uncertainties which she felt existed, the teacher decided, in consultation with some of the parents and with other members of staff, to try sending IMPACT home on a Thursday instead of the traditional Friday. This was in order to give the parents — and/or the children — a chance to come in on the Friday and check out the instructions or have a word about any queries. The teacher felt that it would enable parents to mention any points of uncertainty before they actually found themselves doing the activity with their child over the weekend.

Informing all concerned
Following a discussion with the other staff in the school, the first thing that the teacher did was to get each child to draw a little picture on a note saying 'IMPACT NOW ON THURSDAYS' in black letters.

She and the children had a detailed discussion about the new regime and the reasons for it. The children then took home their picture-note. (In our experience, pictures which are also notes are more likely to be delivered by the children to the parents, and also more likely to be looked at by the parents once they have received them.)

7 Looking at the Effects of Change

Once a change has been made to the routines or practice of IMPACT in a particular class, it might seem a simple matter to see if it has the desired effect or not. Unfortunately, one alteration always involves a whole chain of others, and it is sometimes very difficult, if not impossible, to tell which change is responsible for which effect.

Example. Changing the day. After half a term of the new Thursday sending arrangements, the teacher reviewed the *status quo*. She found that:
- she had more responses than previously,
- she got fewer comments indicating that the parents did not understand the sheet,
- the children seemed to be more relaxed and confident about giving instructions about what had to be done,
- she found that almost no parents took advantage of the new arrangements to 'pop in' on a Friday to sort out any queries.

Some of these results seemed contradictory. For example, why were the parents and children appearing to respond better if they were not taking advantage of the mechanism set up to enable them to respond better?

Obviously, there are a number of hypotheses which could be posited. The parents and children may have been reassured simply by the existence of a mechanism by which they could ask for help. The improvement in responses might be coincidental and have more to do with the particular activities sent than the changes made ... and so on.

Sharing Maths Cultures

What is important here is not necessarily to isolate beyond all reasonable doubt cause and effect. With a process as complex as IMPACT and with this number of people involved at each stage, this is unlikely to be possible. What matters is that we keep notes as to what changes were made and when, and as to what subsequently happened. This means that over the long term we are likely to build up a better picture of what is going on, and ultimately of how to influence it. This is, after all, in part at least, what we are aiming for by evaluating what we do.

Using Evaluation for Critical Reviews of Practice

1 Accepting criticism.
2 Feeling supported.
3 Allowing ourselves to change.
4 Seeing what we have to do.

It is even relatively undisturbing to think about what alterations we might make to particular routines or practices. However, there is an aspect of evaluation which can be uncomfortable, and that is when it becomes more of an evaluation of our own practice, a self-evaluation.

1 Accepting Criticism

Example. An evaluation of IMPACT in a particular school demonstrates that the general level of responses is about 85–90 per cent except in one particular teacher's class, a second-year junior class, where the responses are about 50 per cent.

This immediately presents a very difficult situation. The teacher responsible for this class is going to feel threatened simply by the 'publication' (in the sense of making public) of this information. A feeling of threat is not likely to induce a situation where the issue can be discussed openly or where possible causes can be addressed or remedies suggested.

What frequently happens here is that the situation is either 'buried', i.e. ignored and not spoken of, or it is diffused by only allowing 'safe' explanations to be given; e.g. the parents and/or children in that class are a particularly lazy lot! We are not saying that such 'safe' explanations are necessarily false, but rather that there are other equally plausible explanations which should also be considered.

2 Feeling Supported

The only possible situation in which matters relating to how we are managing our own practice can be discussed is one in which all the participants feel 'safe' in admitting vulnerability and secure in talking about weaknesses as well as competences. How such a situation is brought about will depend upon the school and the nature of the relationships in the staffroom.

Some school staffs or teams of teachers have, through joint planning, team teaching or simply through coping with problems together rather than apart, constructed a safe environment in which difficult questions like this can be addressed and solutions discussed. This is usually achieved through a sharing of ideas and strategies rather than through a more prescriptive approach to the 'failing' teacher. 'When I was having trouble getting some of the parents and children in my class to persevere with IMPACT I found that sending a couple of puzzles, mid-week not at the weekend, really did get them motivated . . .' is the sort of statement which makes a good starting point for discussion. Firstly, it admits explicitly that problems about the number of responses are not unique to one teacher, nor are they once only. They recur, and responses do fluctuate for a whole variety of reasons. Secondly, the statement allows each teacher to make suggestions as to possible changes in practice which might help.

In other schools, such a supportive network is not yet established or, for more institutional reasons, is difficult to set up. In these cases it is often easier to start with the more intimate one-to-one situation where two or three teachers, perhaps those teaching one year group, or doing IMPACT for the first time, can enable each other to explore new ideas and possibilities of change in a non-threatening fashion.

3 Allowing Ourselves to Change

We have found the last five years working on IMPACT to be a very testing time, not only because of the hard work and commitment involved, but mainly because it has been a time in which we have had to cope with some fairly difficult changes in how we work and even in what we think and believe. It is always hard to question assumptions we have held for years, or the value of certain well-established routines or habits. But it is sometimes very necessary, and better ways of working are the result.

Sometimes what the parents have had to say about the ways that teachers are approaching maths has proved more illuminating than we had all, in our prejudices, supposed. In one particular case the whole school staff were convinced that they had the maths curriculum taped. They had all read all the right documents and attended the most recent courses. However, of course, things are never so neat, and when the parents started to be involved on a weekly basis through IMPACT, they had some things to say which threw some of the details of the practice concerned into question. There were some very interesting — and hard — discussions between the teachers and the parents and the teachers found that they were able to evaluate critically certain aspects of what they were doing without automatically throwing into question their whole approach.

On other occasions it has been quite difficult for some groups of parents to accept the fact that there is never only one way to solve a mathematical problem or to approach a particular aspect of maths. For some of us, maths was definitively that subject in which no questions were asked and no answers needed as to the best approach. There was only one correct approach and that was the one that the teacher put on the board! There was an odd sort of comfort in this. At least we all knew where we were.

Teachers nowadays approach the teaching of mathematics in a very different way. Children are encouraged to find — and use — more than one method of achieving an answer. There are many ways of doing any sum and children, like adults, need to be versatile and able to select the appropriate method for the situation. Parents brought up to expect a different approach to teaching and learning maths have had to listen to what the teachers are saying and see what they are doing. This has involved many parents in changing seemingly fundamental assumptions about the best way to teach maths.

4 Seeing What We Have to Do

Sometimes it is not so much a question of allowing ourselves to change as of recognizing where the changes have to be made. If we have been working for a long while in the same way, small routines and habits can become fairly entrenched. Once we have accepted that a new way of working will bring certain advantages, it is still difficult to alter some of the details simply because they go unnoticed. But sometimes it is the fine details, which are generally unremarked, which can make the difference between something working or not working.

Example. Star chart! A teacher may have been used to working in a very formal way with the children, using a fairly rigid and traditional commercial maths scheme and doing very little in the way of practical work. For a while, she/he has felt dissatisfied with the scheme and in general with this type of approach. The children were used to working individually through workbooks and their main motivation was provided by teacher approval expressed explicitly by means of a large star chart on the wall of the class.

The teacher decides to start working using IMPACT, and finds that a number of the sources of dissatisfaction are blocked. The children start to work cooperatively in the work they do in following up their IMPACT. They are more often than not doing some practical maths both in the home and back in school, and they become much more talkative about maths in general. All these are seen by the teacher as very positive changes.

However, the teacher still maintains a star chart on the wall of the classroom. Stars are no longer given to the children for working quickly through workbooks. They are awarded for bringing back their IMPACT. The difficulty here is that the teacher has not paused to consider whether this habit of keeping a chart, which she/he has done for years, is actually a help or a hindrance, in sympathy or out of it with the aims of IMPACT. A

Sharing Maths Cultures

plausible argument is that star charts are devaluing and encourage over-competitiveness. They are devaluing in that they effectively 'tell' children that the action for which the star is awarded is not worth doing in its own right, but rather, is something which one has to be encouraged or bribed to do, and they are very competitive in that they encourage children to measure how they are doing, not against themselves but against the record of others.

The point is not so much that star charts themselves are intrinsically opposed to IMPACT philosophy. It is that the teacher never thought to question this detail of her/his practice. It was something she/he had been doing for years and it was too much of an ingrained habit to be subjected to scrutiny. She/he automatically adapted it to fit the new approaches.

It is these small routines, these invisible habits, which it is sometimes necessary to scrutinize and question, because it is through such details that the innovative practices we may be trying to establish can founder and become unsatisfactory.

National Assessment and IMPACT: an Interim Discussion

Since the Education Reform Act, the role of evaluation has to be seen in the context of assessment in general. National Curriculum assessment is formative and summative, but it is also evaluative. The children are likely to be assessed both over a period of time and also at key reporting ages. Some of these assessments will be reported, and the results will be used to enable parents, teachers, governors and LEAs to make comparative judgements on the basis of these scores. Evaluation of not only children's performances, but of teacher's performances and of school's performances will be partly based on this information. It is not yet clear to what extent schools' reputations and teachers' careers will stand or fall on such information, but it is very clear that it will be an important and newsworthy part of any evaluation of performance.

How then, does this link with what we have been discussing earlier in the chapter about a useful, honest and friendly evaluation of practice in general and IMPACT in particular? We can best see how these two widely different and almost conflicting approaches to evaluation are linked through a consideration of two aspects:
1 Assessment of children.
2 Links with the reporting back procedures.

1 Assessment of Children

In so far as we are concerned on IMPACT to enhance the children's learning of maths, we are concerned to monitor children's performances. How does this link with National Assessment?

We see the assessment of children as an area in which not only teachers are involved but also parents. We involve the parents in trying to assess exactly what their child can do, partly in order that they, as well as the teachers, are aware of how difficult it is to assess.

One of the points that we hope will emerge through a greater partnership between teachers and parents is that it will become clear what an important role assessment plays in the selection of appropriate work for a child. We look at how a child gets along with a particular task precisely in order to make a judgement about what he should do next. It is this aspect of assessment which is highlighted for parents by questions such as:

Do you think your child needs:

a lot more   a little more   no more
    help   help   help

with this subject?

Here, the link between the assessment of what a child can do, and the choice of work for them next, is made explicit.

2 Links with the Reporting Back Procedures

The provisions for the reporting of information outlined in the Education Reform Act are the cause of not inconsiderable anxiety amongst those involved in education. Some of us are fearful that the type and nature of the conversations between parents and teachers which are encouraged or envisaged by Section 22 of ERA, are not those which people engaged in implementing effective home–school partnership would most wish to see.

IMPACT is primarily an initiative in home–school collaboration. On IMPACT, parents and teachers become used to discussing the children's learning within a context in which both are engaged in weekly tasks and activities on which judgements may be based and arguments developed. It is this context which we feel is helpful in a situation in which purely quantitative data is being used for evaluative purposes. If parents and teachers are habitually discussing whether or not the number-line activity was too hard, and the rationale behind the Snakes and Ladders game, they are unlikely to cease to address these questions because of the publication of the aggregated scores of assessment results.

However, the very fact that such results and aggregated scores are readily available will inevitably have an effect upon the nature of home–school dialogue. Parents will want the school to be seen as effective as far as delivering the National Curriculum is concerned. They will clearly be immensely concerned about their own children's assessments. There will be a great deal more comparative data around than has previously been available.

Where a school or a teacher has established ways of working together, this collaboration will set the context within which all the rest of these conversations take place. This can only serve to reduce the potential for misunderstandings or the misuse of assessment data, and decrease any possible conflicts. Sharing activities at home each week does ensure that the parents will have a reasonable idea of what their child can and can't do, including any difficulties she or he is having. These perceptions are based on first-hand experience rather than the results of second-hand assessments. This makes it unlikely that the SATs, or the continuous teacher assessment will throw up any startling information for either parents or teachers. In our experience, world-shattering events are more likely to happen on IMPACT than in the classroom (teachers sometimes complain about this!) and they are more likely to be of the 'Goodness, look what they *can* do!' variety than the other way round.

Afterword

In this book we have raised questions, described IMPACT routines and practices, and discussed a theoretical framework. At times you may have been tempted to pause and wonder, 'Well, I never thought it was all so complicated!' and in a sense, of course, you are right. The basic idea is very simple, and it works.

We may not have conveyed the excitement and the pleasure we have felt when travelling around the country seeing some of the work which parents and children have done, talking to those involved, and hearing the stories. Some of the incidents we have seen or been told of have been enough in themselves to have made the project worth while, and cumulatively, the effect is remarkable.

One teacher sent home an activity that asked which cylinder would hold more — a piece of A4 paper rolled lengthways, or a piece rolled widthways.

The children had to guess in advance, and during the preparatory discussion on the rug, they were all agreed that naturally the two cylinders would hold the same since they were constructed out of the same piece of paper.

One parent's comments went as follows: 'We tried this activity using cornflakes and were very surprised to find that the short fat cylinder held more, since we had all thought that they would hold the same. Thinking cornflakes not a a very good measure (too much air space in between!), we then tried it again using Persil. We still found that the short fat one held more. We cannot work out why this should be so. HELP.... HELP.... HELP....'

We and the teacher were struck by the amount of conversation and discussion (and the mess!) which had obviously taken place. Both the child and the mother could not wait to get into school on the following Monday to discover *why* this was so.

Another child in the class had been equally puzzled. She wrote, 'I was very puzzled at this 'cos I had assumed that they would both hold the same. So I decided to try an experiment. I cut my piece of paper in half, length-

Sharing Maths Cultures

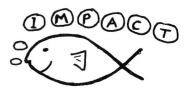

What's in a tube?

You will need a cardboard tube from a kitchen or toilet roll, or a piece of card which you can roll into a cylinder.

Place the cylinder on a plate or dish, or on some newspaper. Fill the cylinder, carefully, with sand, or soil, or rice.

Now empty out the contents, and keep them aside. Now cut the cylinder open, roll it the other way and stick the join with sellotape.
Fill this new cylinder.
Does it hold the same as the other one?

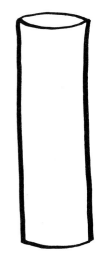

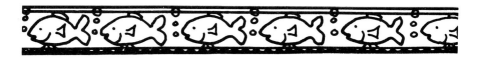

Afterword

ways — and then I sellotaped the two halves together as shown. I then curled this round to make an even shorter fatter cylinder. I found that this one held even more. Then I realized why. It was because the round bit in the middle (which is *not* made of paper!) is bigger, the shorter, the fatter the cylinder.'

We were all fascinated with this 8-year-old child's reasoning. Looked at from one, perhaps bizarre, point of view, she was doing calculus! From any angle, she and her parents had been motivated to discuss and work out a problem in maths in their own terms.

Some of the children's and parent's comments have indicated how much they have both gained from the experience of sharing a particular activity.

'When I first brought the game home, I had to tell my Mum how to play the game. At first, when she saw it, she didn't know what it was all about. Then I told her how it works, and she said it will be easy. And we had good fun playing the game.' (8-year-old)

'We played this game many times during the week, using different totals to capture the planets. We discovered that some totals, i.e. 11, took longer to complete the game, as others such as 7 were much quicker. Louise really enjoyed the dice and devised other little games using them. E.g., she threw the dice, added up the two numbers and wrote the answer in two columns, one for her and one for Benny (her large teddybear). She wanted to do this all the way down the page which would have meant about 50 goes each, so I suggested a smaller number and she said 10 would be better. After 10 goes each, she added up the total — Teddy has 83, Louise has 76.' (Parent)

The responses from parents on these activities are enthusiastic and positive to the point that we would gush with embarrassment to reproduce them here. There will be things that teachers and parents will want to discuss further in the interests of the child's progress. But with such ebullience it is often difficult to do anything other than celebrate in such moments.

This book has been written in a state of tension between providing

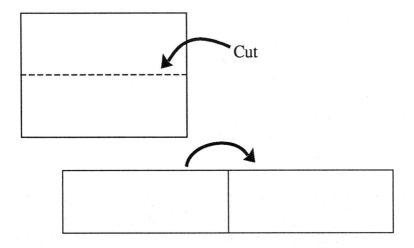

Sharing Maths Cultures

I M P A C T

Name: _____

Did you enjoy the activity?

Why?

If you want, write or draw about what you did.

I M P A C T

Helper's Sheet

Name of Activity _____ Date _____

This activity took ▷ minutes

Who helped the child?

This weeks activity went:

▷ very well   ▷ had problems   ▷ no problems

If you had a problem, what was it?

I worked with my child:

▷ all the time   ▷ most of the time   ▷ some of the time

Please comment on whether your child understood the activity and on how it was completed.

Afterword

IMPACT   IMPACT   IMPACT   IMPACT   IMPACT

CODE <u>MH 3</u>

Activity name <u>Colours</u>   no. <u>2</u>

1. Write your name <u>TERRY</u>

2. Did you like this week's activity?

   not much   it was O.K.   I liked it
      ○   ○   ✓

3. Did you think it was

   too hard   too easy   just right
      ○   ○   ✓

4. If you want, write about what you did..........

   Terry liked blue
   Mummy liked pink.
   The guinea pigs liked green.

Sharing Maths Cultures

IMPACT   IMPACT   IMPACT   IMPACT   IMPACT

CODE Hbl1

Activity name Grab   no. _____

1. Write your name Halil

2. Did you like this week's activity?
   not much   it was O.K.   I liked it ✓

3. Did you think it was
   too hard   too easy   just right ✓

4. If you want, write about what you did..........
   We play the game with my grandmother and with my grandfather. we enjoy so much.

enough in the way of strategy to enable you to try IMPACT and leaving enough open-ended for you to make the project your own. You have a unique understanding of your local conditions and you will have your own ideas. Giving directions can become an involved business especially if you need to get from A to B, but would also like to enjoy the scenery. You might like to consider the following true story.

The authors of this book are often asked to travel to interesting and unusual places to give talks about IMPACT and related issues. On one occasion we were asked to come and talk to a group of advisers in an idyllic convent setting, somewhere in the deciduous depths of the English countryside. We asked for, and received, an excellently laid out set of directions and an eminently readable map. Having eventually arrived at the nearest village, we were comforted to sight the street-lamp, pub and telephone box mentioned in the directions and now beaming reassurance at us. Turning off and now, slavishly, following the map we made the final left turn into the spot marked X. What we saw astonished us. Rather than the gothic haven nestling in the wooded grove we had somewhat fancifully imagined, we found ourselves in a field full of those smells so distinctive to the urban nose. Furthermore, we were in company. A group of other motorists were having an animated discussion accompanied by turbulent gesturing. Yes, they were all advisers, to a man and woman.

Having discarded the idea that, following St Francis, we had been engaged to talk to man and animals in open communion with Nature, we pursued the strategy of driving on in the frenzied hope that by chance we might happen on the convent. And we did happen on the convent. We were greeted by the gentlemen who had sent us directions. He cheerfully pre-empted our suggestions about what he might do with his map by telling us that just a few days before British Telecom had moved the telephone box some fifty paces from its previous location. The timely interjection of the convent bell chastened our animosity into laughter.

Sticking too rigidly to given formulae is an obvious mistake — if one, small but vital element goes missing (or is shifted a few paces) you can end up in places you did not want to be. On the other hand, attention does need to be paid to the essential guidelines that place you in the right vicinity.

Where that 'right vicinity' is for any one of us, who can tell? If IMPACT makes it possible for any of you to, perhaps, renew your commitment, to suggest possible changes in direction, to ask new questions, to reformulate your sense of purpose, to have a better relationship with the parents of your children, to rediscover a lost sense of enquiry, or simply enjoy what you do, we will be more than content.

Appendix

The resources mentioned in this book are all taken from the IMPACT Packs of materials. Details of how to obtain these packs are available from:

The IMPACT Project,
School of Teaching Studies,
Polytechnic of North London,
Marlborough House,
Holloway Road,
London NW1
Tel. 071 607 2789 × 4108.

Details of the IMPACT National Network (INN) and how to become a member can also be obtained from this address.

The activity ideas given in this appendix are sample materials from a number of the IMPACT Resource Packs.

Appendix

ANGLE SEARCH AT HOME

Angles are ways of looking at the amount of <u>turn</u>.

Look around your home for things that turn.

For example,
Door handles
clock hands

Draw at least 5 things which turn.

how much do they turn?

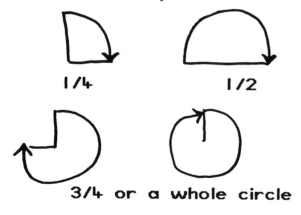

Sharing Maths Cultures

Decimetre Worm

Use your strip of card to make a decimetre worm.
Write your name on one side.
On the other side draw a worm or a caterpillar.
Colour in the caterpillar.

Eg

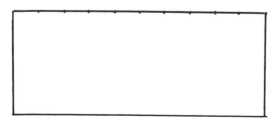

Find a 1p coin.
How many coins can you fit along your decimetre? Guess first. I guess .... coins.
Now try it out.
The decimetre is .... coins long.

Extensions
How much is your decimetre worth? Count up the 1p coins.
....

Now try the same thing with 2p coins or with 5p coins.
................in 2p's   .... in 5p's

Cut out & stick to card.

Appendix

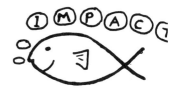

Draw a Map!

Can you draw a map of one place that you have walked to this weekend?

Include things like special buildings, post boxes, sign posts, telephone kiosks, railway lines, bridges, and any other things you notice.

Note to Parents
The children will need a lot of help drawing their maps. Some maps may be very simple and not look much like a "real" map. This does not matter. Encourage children to think about left and right and which way they have to turn.

Eating Up

Think of 3 things which you would like to eat.

Ask someone to write their names and prices. (How much they cost)

Can you draw them?

		Cost
First item	....	
Second item	....	
Third item	....	

Appendix

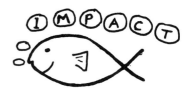

Fraction Activity

Find an adult to help you divide as many of these things as possible in half:

1) a piece of string
2) a glass of water
3) a few ounces of rice or lentils
4) a piece of paper
5) a lump of dough or plasticine
6) a 50p or 20p coin

Can you please bring the two pieces of string back to school?

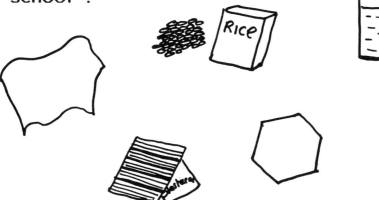

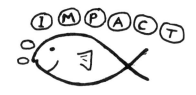

Furniture Count

Look for a wooden piece of furniture to draw.

Count how many pieces of wood have been used to make it.

How many straight pieces ?

How many curved pieces ?

Look carefully at the piece of furniture. Are any of the pieces horizontal or vertical ?

Draw the piece of furniture carefully and bring your drawing into school.

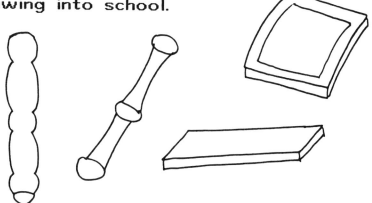

Appendix

MAKE UP NUMBERS

Draw a square.

Choose any number to put into your square.

Put a number at each corner which will add up to your chosen number.

Draw more squares and try some other numbers.

Could you use another shape?

What about a triangle?

Example:

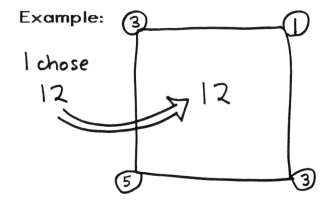

Sharing Maths Cultures

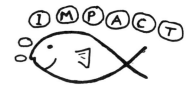

NAME LENGTHS

On the strip of paper attached, ask someone to help you write your name.

You must put <u>one</u> letter in each square.

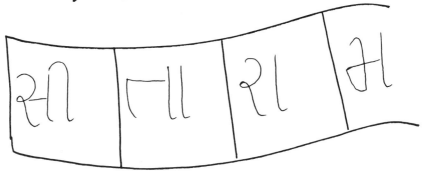

Do the same for someone else in your family.

<u>or</u>

Write your name in more than one language.

Bring your strips into school. You can colour them in first.

Appendix

NEVER-ENDING LINE

Can you draw a long line?

Can it be straight?

Can it be curved or curly?

Now can you draw a never-ending line?

HINT – It might help to think about joining the ends.

What shapes can you draw?

Sharing Maths Cultures

PAPER CHAIN

Use one piece of paper and some glue or paste if you need it.

Make the longest chain that you can. You may cut your paper in any way that you wish.

Bring your chain back to school.

Try cuts like These

HINT! Use old newspaper to try out your ideas!

Appendix

ROCK AROUND THE CLOCK

How many times a day do you have something to eat or drink? Write down the time each time you eat something or drink something.

Also do this for as many people in your family as you can. You could pin a chart on the kitchen wall!
You could try filling it out for the dog or cat....but you will have to watch carefully to see when they have a drink! How about a baby?

Don't forget – every thing you eat or drink counts! You don't have to write down what it is – just when you had it.

Shape Search

Go home, ask if you can use a packet, tin, or bottle from the kitchen.
Stand it on your piece of paper.
Draw round it.

What shape have you drawn?

Try and find another thing with a different shape.
Draw round that.

Draw as many things with different shapes that you can find.

Bring <u>all</u> your drawings into school.

Hint! You could draw round a soap box, a spoon or a banana.

Appendix

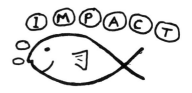

Unit Search

What sort of units are used to tell you how much of something you are buying?

Go round a shop. Look at food.

How many units can you find?

Here are some, can you add others?

Unit	What it's measuring	Comment
gram		
ounce		
pound		
ml.		
cc.		

Sharing Maths Cultures

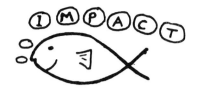

Divide and Rule

Using two straight lines to divide a square, what shapes do you get?

Try as many as you can.
How many shapes are there?
Are there always the same number of shapes?
How many of the shapes can you name?

If you feel adventurous, try three lines...!

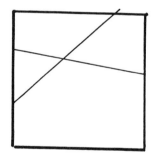

PINT-SPILL!

You will need: string, scissors, and a container which holds exactly one pint.

1) Fill your container with water very carefully.
2) Carry out the full container onto a hard surface (e.g. paving).
3) Lay your string in the shape and size of the puddle, which you think that your pint of water will make when completely spilt.
4) Now spill all of your water into the string! were you right?
5) If you were not right, use a <u>new</u> piece of string to put round the edge of your puddle.
6) Mark your strings to show which is which, and bring them into school please.

<u>WARNING</u>: This activity needs to be done outside.

Sharing Maths Cultures

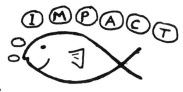

Sum and Product Game

You will need someone to play with. You can use a calculator if you like.

One of you thinks up 2 numbers (you can use decimals).

Multiply them together.............
.... (this is the product).

Add them together ....
.... (this is the sum).

Now tell your partner the product and the sum.

Your partner has to guess what your secret numbers are!

Your partner gives you one to do at the same time.

Who guesses first?

How many numbers did you try?

Appendix

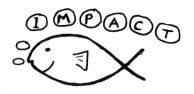

Coin Rubbing

Find three <u>different</u> coins.

Put a little circle of sellotape or piece of blu-tack on a table. Stick your coin onto it.

Now put a piece of paper over the coin and crayon over it <u>gently</u>. You need a wax crayon or a soft pencil.

Rub both sides of 3 coins.

What number do you see ?

Can you write them down and tell someone what they are ?

....

Sharing Maths Cultures

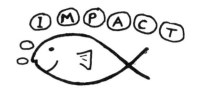

Candle Count

Colour the number of candles you had on your last cake.

Can you work out how many candles you have had on all your cakes ?   .....

In the space below, could you design a very special birthday cake ? (Or make a junk model cake !)

Appendix

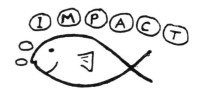

CHANGE OF WEIGHT

Do pasta, rice, or potatoes change weight when you cook them?
Ask your mum or dad if you can weigh either pasta, rice, or potatoes before they cook it sometime this weekend.
weigh the pasta uncooked
      or rice
      or potatoes
Write down the weight _____

After it is cooked weigh it again.
Write down the weight _____

Did the weight change?

Did it loose or gain weight?

What was the difference, if any?

Sharing Maths Cultures

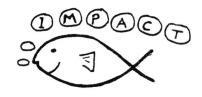

CATAPULTS

Find three different types of paper and make a small ball and a folded pellet with each type.
Use a rubber band to test each ball and pellet.
Estimate how far each ball and pellet has gone.
Check how far each has travelled.

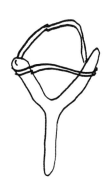

Which goes the furthest?

Which goes the shortest distance?

Record your results
Bring your results and pellets into school.

MAKE SURE YOU DO THIS ACTIVITY SAFELY.

Appendix

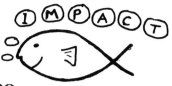

Clocks Around the Home

Look very carefully around your house for as many different things that tell the time.

Draw what you find and put the time that it says.

Put your name on your drawings and bring them into school.

Sharing Maths Cultures

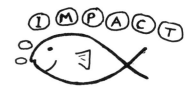

TEAPOTS

Fill your teapot with water.

How many cups do you think it will fill?

Estimate _____

Now find out.
Test _____

Now try different sized cups.

Appendix

Choose One Day

How do you spend your time at home ?

Can you ask a grown-up to help you work out how many hours you spend watching TV, playing and sleeping during one day.

Colour in a block for each hour.

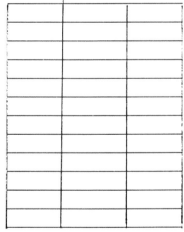

TV/ sleep play
video

Sharing Maths Cultures

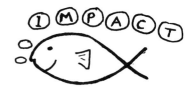

Christmas Boxes

1. Find a box at home.

2. Put inside the box a picture of what you would like for Christmas.

3. Wrap up the box.

4. Bring the box to school as soon as possible.

Notes and Bibliography

1. PLATO. *The Republic*, Book IV — trans. LINDSAY, A. D. (1935) London, J. M. Dent and Sons.
2. The personal opinion of the authors of the present work is that classical subjects deserve a larger proportion of the core curriculum than they currently have and are as relevant to the community and the curriculum as any other subject.
3. BORGES, J. L. (1988) *Labyrinths*, Harmondsworth, Penguin.
4. TAYLOR, T. (Chairman) (1975) *A New Partnership for Our Schools*, A Report of the Committee on School Management and Government, London, HMSO.
5. VASS, J. and MERTTENS, R. (1990) *Delivering the National Curriculum in the Classroom*, London, Heinemann.
6. PLOWDEN, Lady B. (Chairperson) (1967) *Children and their Primary Schools*, Report of the Central Advisory Council for Education, London, HMSO.
7. WOLFENDALE, S. (1983) *Parental Participation in Children's Development and Education*, London, Gorden & Breach.
8. HMI (1978) *Primary Education in England*, A Survey by HM Inspectors of Schools, London, HMSO.
9. THOMAS, N. (1985) *Improving Primary Schools*, A Report of the Committee on Primary Education, London, ILEA.
10. WOLFENDALE, op. cit.
11. TOPPING, K. and WOLFENDALE, S. (1985) *Parental Involvement in Children's Reading*, Beckenham, Croom Helm.
12. GRIFFITHS, A. and HAMILTON, D. (1984) *Parent, Teacher, Child*, London, Methuen.
13. TIZARD, J., HEWISON, J. and SCHOFIELD, W. (1982) 'The Haringey Reading Project — Collaboration between Teachers and Parents in Assisting Children's Reading', *British Journal of Educational Psychology*, **52**, pp. 1–15.

14 HMI (1985) 'Mathematics from 5 to 16'. *Curriculum Matters 3*. An HMI Series, London, HMSO.
15 FROOD, K. (1986) 'Parental involvement in mathematics education; A teacher's view of IMPACT'. *Primary Teaching Studies*, 2, No. 3, London, PNL Press.
16 DONALDSON, M. (1978) *Children's Minds*, London, Fontana.
17 MERTTENS, R. and VASS, J. (1988) 'Special Needs in Mathematics', in ROBINSON, O. and THOMAS, G. (Eds), *Tackling Learning Difficulties — A Whole School Approach*, London, Hodder and Stoughton.
18 DONALDSON, op. cit.
19 DONALDSON, ibid.
20 HUNDEIDE, K. (1985) 'The Tacit Background of Children's Judgements', in WERTSCH, J. (Ed.) *Culture, Communication and Cognition: Vygotskian Perspectives*, Cambridge, Cambridge University Press.
21 BERNSTEIN, B. (1970) 'A Socio-linguistic Approach to Socialisation' in GUMPERS, J. and HYMES, D. (Eds) *Directions in Socio-linguistics*, New York, Holt, Rinehart & Winston.
22 HESS, R. and SHIPMAN, V. (1965) 'Early experience and the socialisation of cognitive modes in children'. *Child Development*, 36, pp. 869–86.
23 SUTTON, A. (1983) 'An introduction to Soviet Developmental Psychology', in MEADOWS, S. (Ed.) *Developing Thinking — Approaches to Children's Cognitive Development*, London, Methuen.
24 VYGOTSKY, L. S. (1962) *Thought and Language*, trans. HANFMANN, E. and VAKAR, G., Massachusetts, MIT Press.
VYGOTSKY, L. S. (1978) *Mind in Society: The Development of Higher Psychological Processes*, Massachusetts, Harvard University Press.
25 WOOD, D. (1978) 'Strategies of problem solving', in UNDERWOOD, G. and STEVENS, R. (Eds) *Aspects of Consciousness*, London, Academic Press.
26 WERTSCH, J. V. (1985) *Culture, Communication and Cognition; Vygotskian Perspectives*, Cambridge, Cambridge University Press.
27 BROWN, A. and FERRARA, R. (1985) 'Diagnosing Zones of Proximal Development', in WERTSCH, J. ibid.
28 GLACHAN, M. and LIGHT, P. (1982) 'Peer Interaction and Learning: Can Two Wrongs Make a Right?' in BUTTERWORTH, G. and LIGHT, P. (Eds) *Social Cognition: Studies of the Development of Understanding*, Brighton, Harvester.
29 DOISE, W. (1978) *Groups and Individuals*, Cambridge, Cambridge University Press.
30 SHOTTER, J. (1984) *Social Accountability and Selfhood*, Oxford, Basil Blackwell. (This is a book of collected works which discusses the social context of 'cognition' in some detail, but unlike other books on the subject also examines the character of the 'social' in some depth.)
31 VASS, J. and MERTTENS, R. (1987) 'The Cultural Mediation and Determination of Intuitive Knowledge and Cognitive Development', in

MJAAVATN, E. (Ed.) *Growing into a Modern World*, Trondheim, Norwegian Centre for Child Research.
32 PERRET-CLERMONT, A., BRUN, J., SAADA, E. H., and SCHUBAUER-LEONI, M. (1984) 'Learning: a Social Actualisation and Reconstruction of Knowledge', in TAIFEL, H. (Ed.) *The Social Dimension of Learning*, Massachusetts, Cambridge University Press.
33 WOOD, D. (1986) 'Aspects of Teaching and Learning', in LIGHT, P. and RICHARDS, M. (Eds) *Children of Social Worlds*, Cambridge, Polity Press.
WOOD, D. (1988) *How Children Think and Learn*, Oxford, Basil Blackwell. (A fuller and broader treatment of the psychology of thinking and learning in children than we can provide in the present work: devotes a chapter to mathematics.)
34 TOPPING, K. (1988) *The Peer Tutoring Handbook*, Beckenham, Croom Helm.
35 GOODLAD, S. and HIRST, B. *Peer Tutoring: A Guide to Learning by Teaching*, London, Kogan Page.
36 WELLS, G. (1986) *The Meaning Makers: Children Learning Language and Using Language to Learn*, London, Hodder & Stoughton.
37 GRIFFITHS, A. and HAMILTON, D. (1984) *PACT: A Handbook for Teachers*, London, Hackney Teachers' Centre.
BRANSTON, P. and PROVIS, M. (1986) *CAPER: Children and Parents Enjoying Reading*, London, Hodder & Stoughton.
38 MERTTENS, R. and VASS, J. (1987) 'Raising Money or Raising Standards?' in *Education 3–13*, June 1987.
39 RANSAC (1989) Records of Achievement National Steering Committee Report, London, DES.
40 HMI (1983) *9–13 Middle Schools, An Illustrative Survey*, London, HMSO.
HMI (1985) *Education 8–12 in Combined and Middle Schools*, London, HMSO.
41 PYE, J. (1988) *Invisible Children*, Oxford, Oxford University Press.
42 GRIFFITHS, A. and HAMILTON, D. op. cit.
43 TOPPING, K. op. cit.
44 WOOD, D., MCMAHON, L., CRANSTOUN, Y. (1980) *Working With Under Fives*, London, Grant McIntyre.
45 STUBBS, M., ROBINSON, B. and TWITE, S. (1979) *Observing Classroom Language*, Milton Keynes, Open University Press.

LIBRARY
WESTMINSTER COLLEGE
OXFORD, OX2 9AT